Science Safety in the Community College

Science Safety in the Community College

By John Summers, Juliana Texley, and Terry Kwan

NSTApress
NATIONAL SCIENCE TEACHERS ASSOCIATION
Arlington, Virginia

NSTApress®
NATIONAL SCIENCE TEACHERS ASSOCIATION

Claire Reinburg, Director
Judy Cusick, Senior Editor
Andrew Cocke, Associate Editor
Betty Smith, Associate Editor
Robin Allan, Book Acquisitions Coordinator

ART AND DESIGN Will Thomas, Jr., Director
Shennen Bersani, cover art
Equipment for cover art provided by Bio-Rad Laboratories
Linda Olliver, interior illustration

PRINTING AND PRODUCTION Catherine Lorrain, Director
Nguyet Tran, Assistant Production Manager
Jack Parker, Electronic Prepress Technician

NATIONAL SCIENCE TEACHERS ASSOCIATION
Gerald F. Wheeler, Executive Director
David Beacom, Publisher

09 08 07 06 4 3 2 1

Library of Congress Cataloging-in-Publication Data
Summers, John, 1936-
Science safety in the community college / by John Summers, Juliana Texley, and Terry Kwan.
p. cm.
Includes bibliographical references and index.
ISBN-13: 978-0-87355-271-4
ISBN-10: 0-87355-271-7
1. Science--Study and teaching (Higher)--United States. 2. Science rooms and equipment--Safety measures. 3. Laboratories--Safety measures. I. Texley, Juliana. II. Kwan, Terry. III. Title.
Q183.3.A1S93 2006
507.1'173--dc22

2006007834

SCILINKS. THE WORLD'S A CLICK AWAY

Featuring SciLinks —a way of connecting text and the internet. Up-to-the-minute online content, classroom ideas, and other materials are just a click away.

Contents

Introduction

Community colleges form the backbone of our nation's system of higher education. More than 1,650 institutions across the nation serve a broad range of students, providing academic, social, career, and citizenship skills for a lifetime of learning.

Community college students are among the most diverse groups of learners in the country. They come from a wide variety of secondary institutions—or none at all, because GED (General Educational Development) diploma holders and home schoolers are as welcome as the local high school valedictorian. New Americans and grandparents sit next to precocious secondary students trying to get a jump on their college studies. Students ranging from those who have had the structured support of special education to those who have experienced no structure all meet in the community college classroom.

Engaging in real science isn't just an academic ideal: It is essential to the understanding of evidence-based reasoning. But the diversity among students in an introductory community college science course makes creating a safe laboratory environment a tremendous challenge. When everyone is invited to participate without prerequisites, nothing can be assumed. The community college instructor can never rely on a common foundation of understanding or skills. This includes laboratory skills and safety precautions. Teach the basics. Do not assume prior knowledge and experience.

Challenging? Of course. But when we teach adults the skills they need to inquire about real-world scientific issues in a safe environment, they gain attitudes they will carry with them to their homes, their families, their workplaces, and their neighborhoods.

We believe that every adult in this nation should be scientifically literate and that we would have a stronger nation because of it. That means there's no place where promoting scientific literacy is more important than in the introductory laboratory classes offered at a community college. We also believe that with background knowledge and good sense, every community college instructor, full time or adjunct, can implement a fully investigative laboratory science course in a safe learning environment. To do so requires planning and preparation, but it's well worth the effort. So it is with great enthusiasm that we offer this guide to safety in the community college.

This book is one in a series of four that offer positive options, even as instructors learn more about hazards: *Exploring Safely* is the elementary school guide, *Inquiring Safely* is the middle school volume, and *Investigating Safely* is the high school guide. In *Science Safety in the Community College*, most of the guidelines for programs, coursework, and student behaviors have been formulated with the introductory student in mind. As students progress through a college sequence, they increase their knowledge and skills, and the standard of best practice changes as well. We expect instructors will use their experience, training, and judgment to decide what might be reasonable for more advanced students. However, the guidelines for facilities—particularly those regarding space, safety equipment, and ventilation—should be considered standard for all course work, whatever the level.

Although the traditional safety manual is often a compilation of safety rules, regulations, and lists, this book takes another path. We offer a more narrative style, providing discussions of safety concepts in the context of commonplace situations in real classrooms. As we did in the first three volumes, we've included many anecdotes that highlight and reinforce ideas. We have changed names and made other modifications to better illustrate some hazards, but all of the stories are based on actual events. We hope this approach makes the book enjoyable to read as well as valuable to reference.

Because we recognize that another way to use the book is to look for specific topics, we have included a detailed index. You will find that some of the same information is repeated in several sections. This is to minimize flipping back and forth among chapters. A glossary at the end of the book defines terms as we use them and includes definitions of acronyms.

We hope the books are thought provoking. No single publication can cover every possibility or all the specific policies and rules promulgated by federal, state, and local authorities. Each state's community college system is different, and the buildings in which the programs are housed differ widely. Our goal is to provide you, the instructor, with examples of safer practices and to help you become more alert to ways of ensuring safety when you teach science in your laboratories and in field studies. Above all, we encourage you to use common sense and stay up-to-date with policies and regulations.

We believe that creating a safe environment for laboratory instruction in the sciences is a group endeavor, led by the instructor but joined by the entire institution. We have included information we hope you will share with department chairs, facilities directors, administrators, and others so they fully understand the support they must provide to enable you to conduct a safe and effective program. As you read, we hope this book helps you "see" your physical environment and your procedures through a safety-conscious lens. That will help you give your students habits of mind to last a lifetime.

No safety book can be omniscient. No guide can prescribe every action or precaution that might be needed in a college classroom. Safety is more than a set of rules: It's a state of mind. Because no single volume can anticipate everything that could go wrong, we believe instructors and administrators must make awareness of safety issues a skill and a habit. So we hope that as we share these experiences with you, you'll hone your own safety skills and that sixth sense that creates a stimulating and safe classroom for all students.

Acknowledgments

Thanks to Betty Smith, our editor at NSTA, and to the contributors who provided advice and reviewed and added to this document: Ken Roy, Parris Powers, James Kaufman, Sandra West Moody, Jennifer Fischer-Mueller, Nancy Lane, and Fred Wang, MD. Their tireless work has helped us polish our view of the classroom and enrich our offerings to you, the reader.

The authors have been working together for many years as part of the NSTA TAPESTRY grant program funded by Toyota Motor Sales, USA, Inc., and wish to acknowledge with thanks the generosity and support that Toyota has provided to hundreds of science educators and thousands of their students for more than a decade.

Author Biographies

John Summers taught environmental sciences, biology, and chemistry for many years and continues to be involved in programs to support teaching and learning of the sciences at the precollege level. A presenter at numerous NSTA and American Association for the Advancement of Science (AAAS) conferences, he is also a faculty member for online teacher training at the undergraduate and graduate levels. He has served on panels to structure and review frameworks, assessments, and systemic initiatives in the state of Washington. His special interests include using science-oriented outdoor experiences to challenge and connect with at-risk students.

Juliana Texley has taught all the sciences, K to graduate school, for 30 years and spent 9 years as a school superintendent in Michigan. For 12 years she was editor of *The Science Teacher,* NSTA's journal for high school instructors, and served as an officer of the Association of Presidential Awardees in Science Teaching. She currently teaches college biology and technology and develops and teaches online courses for students and instructors.

Terry Kwan taught middle school science before becoming a science supervisor and teacher trainer. She is an independent contractor, collaborating with private and public institutions to develop science programs, train teachers, and design science facilities. She served 18 years as an elected school board member in Brookline, Massachusetts, and currently serves as a lay member of the National Institutes of Health Recombinant DNA Advisory Committee and a community representative to Institutional Biosafety Committees for the Harvard Medical School and the Dana-Farber Cancer Institute.

How can you and your students avoid searching hundreds of science websites to find the best information on a topic? SciLinks, created and maintained by the National Science Teachers Association (NSTA), has the answer.

In a SciLinked text, such as this one, you'll find a logo and keyword in the text near a concept your class is studying, a URL (*www.scilinks.org*), and a keyword code. Simply go to the SciLinks website—*www.scilinks.org*—type in the code, and receive an annotated listing of web pages that have been extensively reviewed by a team of science educators. SciLinks is your best source of pertinent, trustworthy internet links on subjects from astronomy to zoology.

Need more information? Take a tour—*www.scilinks.org/tour.*

1

Setting the Scene

Safer Science in a Drive-Through Learning Community

More than 1,650 community colleges across the country provide convenient, close-to-home access to learning. But the convenience and broad appeal of community college courses also create challenges for the instructor. Many community college students are distracted by competing obligations—family, job, and other responsibilities may vie for time and attention. Because students don't have to pay high fees or move away from home, they may treat introductory courses like an academic smorgasbord. Some may come early, ready to enjoy a feast. Some may stop in to sample the menu, ready to drop a course at the first sign of challenge. Still others transfer in when a different course gets tough, hoping that a science course is more to their liking. The challenge is to keep every guest at the science banquet and enable each to develop a true appetite for science inquiry as a critical skill in comprehending the world. Along with the challenge, community college instructors may have a unique advantage. Unlike teachers of high school or four-year college students, community college instructors can count on a powerful characteristic in most of their students—motivation. Community college participation is almost always a voluntary act, motivated by the student's personal goals. It takes a special effort to drive to each class on time, prepared and ready to work. Though you may compete for their attention, your students want and need what you have to offer. You hold the keys to their future.

A Scientifically Literate Citizenry

The National Science Education Standards (NSES) (NRC 1996) leave no room for doubt—inquiry-based science is vital to producing scientifically literate adults. Although they were written primarily for K–12 programs, the critical thinking and analytic skills inherent in scientific literacy are as important to all adults as to professional scientists. The same skills and thought processes engendered by laboratory investigations are those your students will need in the workplace, in the home, and in the voting booth. This way of knowing is critical for successful citizens and to a functional society.

Even if you represent a student's one and only exposure to science, your role is vital. The introductory science course may be the last opportunity members of our increasingly diverse society have to learn how to gather, evaluate, and draw conclusions from empirical evidence. What they learn or do not learn can influence significantly the way they make decisions as citizens, employees, managers and leaders, parents, and voters.

Evaluating choices and accepting, rejecting, or modifying recommendations based on data occur regularly in adult life whether a student chooses a career in science and technology or not. An individual's ability to weigh options with scientific and logical rigor, to reject pseudo-science, or to accept scientific analyses may well be based on that person's experiences in the course you teach. Providing a program that demystifies science and puts it in a safe and understandable context may be even more important for students not planning to pursue careers in science and technology than for those who do.

Moving Up Safely

Developing a responsible and safe introductory community college laboratory science program is a challenge. The subject matter is complex, requiring cerebral, technical, and mechanical skills. The prior knowledge and experiences of students are diverse—they range from retired professionals returning for intellectual stimulation to high school dropouts who have discovered the need for education and just passed their General Educational Development exams.

For most instructors, what can be controlled is limited. The course outline, the prerequisite requirements for students, and the physical facilities often are prescribed institutionally or by groups of instructors. In larger community colleges, the responsibilities for parallel sections may be shared by full-time and adjunct instructors. They must share facilities, equipment, and supplies even though they rarely meet. Although community college students may receive information through course catalogs, they make decisions regarding registration and attendance with much less guidance, structure, and support than afforded to secondary students.

These circumstances make it all the more important that you, your colleagues, and administrators have structured guidance to ensure that laboratory investigations have been selected and designed with safe practice in mind and conducted in facilities that are appropriate. We hope this book provides some of that guidance and that, more important, it reminds all involved that specific attention must be paid to safety for all laboratory science instruction.

Using This Book

After you read this overview chapter, the way you use this book will depend upon your experience, your training, and your assignment. You'll find the book includes a lot more than just a list of safety rules. There are tips for teaching, management, and even your own mental health. You may wish to skip right to your own discipline (in Chapters 5 through 8) or use the index to find best practice on a specific topic, such as Standard (Universal) Precautions, contact lenses, or internet safety. Or you may want to use the book as a refresher for your own professional development, reading it from start to finish.

If you choose the latter course, you'll find some repetition. Some ideas deserve to be repeated. Some appear more than once to accommodate those who read only a part of the book, including administrators, support staff, and disciplinary specialists. Other ideas may seem obvious to full-time veteran faculty but may surprise adjuncts who come to teaching from business or industry.

No matter how you choose to read this book, you are likely to find something that makes you uncomfortable—a treasured demonstration or customary practice that isn't considered safe. You're not alone. We've had the same experience in writing and working with instructors while developing the book. Change is never easy. You'll likely find some precautions that you just don't believe; your experience and training may have convinced you that they are not only safe but also essential to your goals as an instructor. That may be true and may be appropriate in your program. But it's also possible that thus far you've just been lucky. So we invite each of you, our peers, to consider seriously the precautions we've included and to use good judgment in adapting them to your assignment and your institution.

You're likely to find differences between this book and the traditional safety manual. Our recommendations are based not only on physical dangers but also on developmental and experiential appropriateness. In many cases, we don't discuss just *how* to make an investigation safe, but also *why* it should or should not be done at a given instructional level. This may be a new perspective for many veteran instructors, but we have found that laboratory accidents can frequently be traced to assuming that students bring knowledge or experiences that the instructor or former students have had, but which the current population has not had. In this day of butane lighters and piezoelectric sparkers, you may be quite surprised at how many students don't know the difference between a safety match and an ordinary friction match. You may be quite familiar with sterile technique, but your students may not have studied—or may have totally forgotten—germ theory.

Chapters in Brief

The low cost and convenience of community colleges are two factors that draw the diverse population who attend. Among the things students are looking for at a community college are a slow transition to college work, remediation, and "cover" to stay at home or in the country. What is almost always a common factor is the proximity of the college to the students' residence or workplace. Unlike high school classes, which are frequently leveled by interest and ability, or four-year college and graduate courses

1

in which students are sorted by majors and prerequisites, the community college class is likely to have students with a plethora of interests, prior experiences, and goals. In Chapter 2, "Communities of Learners," we look at ways to accommodate diverse needs and goals—not only to make your program safer but also to help everyone involved achieve the goals that brought them to the community college campus.

Many of our college buildings are aging. In some institutions, increases in enrollment have forced administrators to use stopgap scheduling or emergency housing. National studies have found a significant increase in the number of buildings in serious need of repair, while shortfalls in state budgets have left publicly supported colleges with few extra dollars for renovation or expansion. Facilities are vital to safe science, as you will see in Chapter 3, "Where Science Happens," which discusses space and equipment, and Chapter 4, "Finders Keepers," which discusses storage.

Meanwhile, the body of content knowledge in each discipline of science constantly changes. What was yesterday's best practice can be today's unacceptable risk. Chapters 5 through 8 summarize many of the most common cautions in the science disciplines for instructors in life, Earth, and physical science classrooms. In Chapter 9, we've included tips for field studies and exploring "The Great Outdoors."

The 21st century has brought new technology as well as new concerns. After covering traditional safety categories, we added Chapter 10, "The Kitchen Sink," which discusses items that fit nowhere else, such as internet use, allergies, and Standard (Universal) Precautions.

Today's society is quite litigious, and many people place blame on instructors for factors well beyond the instructors' control. So finally, in Chapter 11, we share tips on how professionals can "Live Long and Prosper" by protecting themselves as well as their students.

Keeping Up

What do those degrees on your wall mean? A prescribed sequence of courses? A statement of competency? A ticket to a job? Since your last formal course work, a great deal could have changed, both in your science discipline and in teaching techniques. An important step to safer college science is to keep on learning.

It's not just the content of science that changes but also the standards for safe investigation. To structure a safe college science program, you must stay current with both research findings and regulations. Which chemicals now require special handling or are banned outright? What cultures and culturing practices previously thought harmless are associated with serious infectious disease? What are the latest requirements for protective gear? Best practice is constantly changing. You must remain up-to-date and so must your course outlines. That's especially challenging for a faculty that is partly or predominantly adjunct.

Students change too. Their interests and developmental levels are affected by previous school experiences and by their experiences outside school. Demonstrations once considered motivational—such as explosions—are now considered dangerously tempting to students who have been bombarded by media violence. It's not enough to simply evaluate the content of an inquiry; you must also consider the context in which today's students will receive it.

The information that qualified you for your degree and the course outlines and lab assignments you first used may quickly become outdated. You can't depend even on a book like this one for your entire safety knowledge base. Each disciplinary expert should assume responsibility for keeping up-to-date in his or her area. To help you do that, internet links and references accompany each chapter. We've also included information that leads you to online course work and mentoring by professional associations. We hope you'll join us as lifelong learners in the important endeavor of ensuring safety in science investigations.

It May Take a Village

Some battles have to be fought by the department rather than by individual instructors. For instance, although science instructors may be knowledgeable in multiple science disciplines, it's usually unwise to assign someone to instruct in more than two distinctly different laboratory courses in any single term. The challenge of keeping up with rapidly changing standards and best practice is just too great.

Late registration is another issue. Instructors often begin the semester with a course overview and a review of basic safety procedures. The first lecture session is the opportunity to make sure students understand the syllabus, and the first laboratory session usually is used to introduce students to laboratory facilities and general expectations for lab work. If a student registers late and shows up after these presentations, you are still responsible for ensuring the student receives the same introductory instruction—especially as it relates to safe laboratory practice.

It may be hard to convince college registrars to turn down the prospective revenue of a late registrant. But it is essential that everyone understand and work to mitigate the risks. It may be that the department needs to create a separate laboratory introduction session or exercise that must be successfully completed by all students. This might be incorporated in a web-based tutorial on an online learning platform, such as Blackboard or WebCT, that can record and document successful completion by each individual. It must be thorough, and documentation of successful completion must be non-negotiable.

Safe, well-organized, clean laboratory space is also a condition that can be achieved only if all instructors participate. Community college laboratories are likely shared by multiple instructors and often have no individual clearly responsible. Safety equipment

1

SAFETY IN THE SYLLABUS

Consider where these key ideas might be built into your course syllabus:

- Clearly describe your goals and expectations. Specify not only content, but also laboratory skills, practices, attitudes, and conduct.
- Make students responsible for regular on-time attendance and for making up all instruction missed because of lateness or absence.
- Establish that students who miss preparatory and safety instructions will not be permitted to participate in laboratory work until all preparatory and safety work is made up.
- Identify all the ways you will be offering safety and lab practice guidelines: orally, in writing, on a web learning platform, or some other format.
- Provide specific options for students with disabilities, those with limited English proficiency, and those with limited access to technology.
- Include specific steps you will take to measure student understanding of safety issues and laboratory practices prior to activities where such knowledge is necessary. Do not depend solely on distribution of handouts and head nods to indicate understanding of critical safety issues, especially for students with disabilities or limited English proficiency.
- Insist on students' presence from beginning to end. Make it clear that critical instruction will begin at the start of class, and that latecomers may be excluded.
- Set standards for how quickly—or slowly—you expect work to be done. Avoid using a system where students can race through a laboratory exercise and leave early. This rewards speed and often engenders carelessness.
- Clarify how a student can make up laboratory work—if at all.
- Incorporate a rubric for safety, cleanliness, and organization in the grading scale.
- Affirm that each student is responsible for being an active participant in the education process and ensuring safe, appropriate laboratory practices at all times.

1

may be missing, outdated, or untested; hazardous materials may accumulate; and valuable or potentially dangerous equipment or materials may be left unsecured. Common expectations are needed so that instructors who may never meet each other can expect that certain minimal standards are applied.

You *can* fight city hall—or the dean's office—if you act collectively. It's easiest if everyone is on the same team. If your department head or administration is unaware of ever-changing regulations, safety risks associated with late enrollees, minimal standards for laboratory facilities and equipment, and the need for regular, licensed hazardous waste removal, you may need to tell them. If all else fails, share some of the guidance and legal precedents that we've included in this book.

Planning for the Safe Classroom

The paragraph describing a course in the college guide is usually written by committee. This is not always true for advanced courses, but it's important that all students in a college get the same basic knowledge and skills in their introductory science experiences. That means that an individual instructor may not have much influence on the profile of a specific course. But what if the activities that a committee—or your predecessor—has chosen don't seem safe? No matter what the history or how great the institutional pressure, it is important to remember that the students and sequence assigned under your name are your responsibility. In case of an accident, saying "It was always done that way" just won't fly.

Because each member of a department may have a unique schedule and because meeting time may be scarce, many schools have established online discussion forums for their faculty. These opportunities for almost real-time shoptalk are invaluable. Even if you are assigned only one course on an occasional basis, make sure you participate all year long. Your input—and the choices you make—may depend upon every member of the faculty having the same current knowledge of risks and responsibilities.

The Syllabus Is the Thing

Many schools make syllabus preparation easy by offering templates so they can be quickly prepared. You may be provided with older course outlines that have been used for decades and be tempted to merely tweak the details. Or, you may be given a course syllabus prepared more recently by someone else. In today's litigious society, we strongly recommend that you consider the syllabus you hand out to your students as a written contract between you and your students and that you incorporate language that clearly sets out your expectations and student responsibilities—even if that means providing an addendum to the existing document. Think of the syllabus as written evidence that you have provided a well-thought-out description for the conduct of the course and the conduct of your students.

1

The Best-Laid Plans

You will lower your stress and your liability, whether you are an experienced full-time instructor or an adjunct, if you establish the discipline of preparing complete and detailed written plans. As you'll see in Chapter 11, these plans not only create peace of mind but also provide valuable documentation in case a problem occurs. But recognize that detailed lesson plans can look great on paper and fall short in practice. The best format is one that is convenient for you to prepare and easy for you to follow. You may want to make columns in a plan book to list materials to prepare, time requirements, chemical allocations, and safety reminders.

An important planning principle is that old lesson plans should not be reused without examination and evaluation. Even with revisions, do not keep a plan beyond a set time—say, two or three years. When that time has elapsed, develop something new to ensure that your material is current and your teaching is fresh and enthusiastic.

Make Every Minute Count

If you've had a strong background in teaching strategies, you may be familiar with the research on how the brain stores information. If you've come to education from research or industry, this may surprise you—students don't concentrate much longer in minutes than their age in years, and that linear progression levels off at the age of 20. College students' attention, regardless of age, begins to wander if you try to talk for more than 20 minutes without a break or change in activity.

Using 20-minute elements is key to planning a college science block—especially for courses that meet for several hours once or twice a week. Attention span is also a factor to consider in safety instruction. Plan your sessions as a series of short, varying activities that each last no more than 20 minutes. Combine didactic instruction with debate and discussions, large-group instruction with study group or laboratory group work, safety instruction

Dampening the Swinging Door

A problem for some community college students is consistent and on-time attendance. Emphasize your expectations for consistent attendance in a number of ways:

- Begin classes with a short activity that requires a specific product or outcome. Assign a small percentage of the course grade for these activities. (This is a great way to review safety.)
- Provide examples of appropriate and inappropriate reasons for absence. While you should be supportive of major life events, emphasize your high expectations for regular class attendance.
- Make it clear in the syllabus that failure to be present for the safety instructions will result in exclusion from the laboratories.
- Set firm rules for cell phone use. Under ordinary circumstances, cell phones should be turned off before a student enters the lab. For the rare instance when there must be instant contact with a student, silent paging should be in use.
- Assign responsibility and credit for group interaction. Make members of a group responsible for instructing and reinforcing procedures to all members.

with actual use of equipment. This approach will make your class more interesting, more effective, and safer.

You'll find some specific ideas for session elements in this book in sections called "The Savvy Science Instructor" at the end of each chapter. You can find many more ideas in the National Science Teacher Association's (NSTA) journals available online at *www.nsta.org*.

Never Assume

Some students walk into your classroom with cell phones and iPods but have never seen or used a manual can opener or tuned a radio without a scan button. Others may bring years of life experience to their college science course but seek a quiet corner in the back of the room, intimidated by the energy of the younger students and hesitant to apply highly developed life skills in the academic environment. Learning-disabled students, often supported by special education programs through most of their prior education, may be working without this support for the first time. English language learner (ELL) students may nod frequently, leading you to believe they understand all of your instructions, but those nods may reflect their respect for your position or their desire to be cooperative rather than true comprehension. The first safety guideline for instructors in an intro course is "never assume." Special steps may be necessary to ensure everyone has really grasped the critical information you are trying to convey.

As you introduce your students to college science, try to see your program through the widely differing experiences—or lack of experience—of all your students. Think of your laboratory facility as an entirely new venue and laboratory investigation as an entirely new experience. Do not assume that a student has experienced previously any skill or procedure—not even how to safely handle matches.

LABORATORY ASSISTANTS

One of the luxuries of most colleges is the help that instructors get from paid or credited laboratory assistants. Most lab assistants are supervised by a coordinator and specially trained in safety. Their access to chemicals and other potentially dangerous supplies is often controlled by strict institutional policies.

But no matter how well trained your laboratory assistant and no matter how well supervised, the ultimate responsibility for safety in your laboratory rests with you, your actions, and your instructions. There are many tips

(cont. on next page)

(cont. from preceding page)

for lab management in the chapters that follow. But in general, when you work with laboratory assistants:

- Know and test your laboratory activities thoroughly. Make sure you have tried every part of every lab in advance and are aware of necessary precautions.
- Check all institution, local, state, and federal regulations that might apply.
- Ensure that you and your laboratory assistants model appropriate safety precautions, including use of eye protection and other safety gear.
- If your assistant has the authority to decant and dispense individual quantities of chemicals for labs, check out the chemicals that have been dispensed. Make sure that all reagents are fully and correctly labeled, that quantities are appropriate and correct, and that all stock bottles have been correctly re-secured.
- Personally check and test all equipment you will be using or needing (including all safety equipment) before the laboratory activity begins.
- Ensure that you have the MSDS (see Chapter 4 for MSDS details) on hand for all materials to be used.

Mix or Match: Safety Is Still a Must

In some college programs, each lecture (or content) section is matched with a required laboratory component. In others, the laboratories are sequenced independent of the lectures. When the two sections are matched and taught by the same instructor, there are many opportunities to reinforce key ideas and to build the rapport that supports responsible conduct.

When labs are scheduled as independent units, the basic rule of "never assume" becomes even more important. Students who come into these sections may seem knowledgeable but lack important concepts that would enable them to recognize and avoid dangers. The independent laboratory almost always needs its own built-in series of direct instruction sessions to make sure everyone is thoroughly informed.

The Teachable Moment

Instructors commonly begin laboratory science courses with a tour of the facilities to familiarize students with the location and function of safety and emergency equipment such as fire extinguishers, the safety shower, the fume hood, and the eyewash, followed by a review of general safety rules, such as when protective equipment must be used. They often then give a safety test and keep the results as a record of the session. There's nothing wrong with this practice, but it may be insufficient to ensure safe practice by students. Safety instruction that is abstract and isolated from the activities to which the instruction applies is easily forgotten and will need reinforcement.

You should present general safety procedures—such as using eye protection, decontamination, and hand washing—at the beginning of the course. You also should repeat specific safety instruction in conjunction with the lesson or activity for which the safety procedure is needed. Even if you have reviewed the procedure a number of times, go over it *again* each time your planned activity requires the precaution. Keep a dated record of the specific instruction given to students who were absent or who later transferred into your course. Ensure that every student has received introductory safety lessons as well as the safety instructions associated with each investigation, and document the fact in a plan book or calendar.

Say It Again, Sam

With a revolving door on their lab courses and a diverse student body, community college instructors must be especially vigilant that each student has received all necessary safety instruction. Providing students with a written version of your instructions and safety directions and repeating them at the beginning of each relevant activity is important. When students are absent, they may miss safety directions. Keep an explicit record of what safety instruction has been given, when, and to which students. Keep this checklist as evidence that you gave proper and appropriate safety information to each student. Some instructors prepare and maintain a spreadsheet identifying students and safety issues. When a safety item has been presented, the appropriate square can be dated, providing a dated record of each student's safety preparation.

Some community colleges have open-enrollment policies that permit enrollment by students with only minimal English fluency. Depending upon institutional policy and your own good judgment, you may also need to offer safety instructions in other languages. Instructions may have to be tailored to students with specific learning challenges. You will also need to include appropriate methods of assessing comprehension. For more on this issue, see Chapter 2.

Out-of-Class Assignments

You are responsible and can be held liable for your out-of-class assignments. This includes independent study work and projects. Consider assignments carefully to avoid placing students in hazardous situations. Long work hours and family responsibilities may force students to squeeze homework into spare moments in random places, so consider your students' schedules and other obligations when assigning homework.

1

Try to structure assignments that have practical application and that can incorporate students' work, home, and other everyday experiences. For example, when instructing in microbiology, ask students to relate concepts to regulations and practices at their food-handling jobs or to propose appropriate hygiene practices in changing diapers. In chemistry classes, have students relate chemical characteristics with home or workplace chemical handling and hazardous waste practices. When introducing the concepts of variables and controls, ask students to apply these concepts to news reports regarding drug safety and FDA recommendations.

A Reputation for Excellence—Instructor as Model

Your classroom should be a professional place where everyone is expected to work seriously and dress appropriately. This is both for safety and for career preparation.

Model appropriate dress in your own attire—nothing baggy, torn, or hanging. Lab coats may be part of the caricature of the mad scientist, but they are also part of safe science, especially when you might be working with biological materials, stains, or toxins. Always use the appropriate eye protection during your own demonstrations and whenever you require your students to wear eye protection.

No food or drink in the lab applies to instructors as well as students—and no coffee or snacks, even in the back room if that back room is used for preps. Chemical and specimen-storage refrigerators must have prominent signage indicating "no edible food storage" and must never be used for lunches or other items meant to be eaten.

The physical environment also can contribute to or distract from the quality of instruction. Classroom clutter that cannot be distinguished from a midden both detracts from the seriousness of your endeavor and poses unnecessary safety hazards—fire, tripping, and undetected theft of valuable or hazardous materials.

Set High Expectations

As any veteran instructor knows, high achievement is the reward for setting high expectations for students. This is as true for safety as for any other expectation. The more you make students responsible for using and enforcing safe laboratory and fieldwork procedures, the more easily safe practice can become habit.

College science offers a rich curricular framework within which to build a safe environment for investigation. Safe work habits developed in the college laboratory should engender safe practices used at home and at work. With help from peers, friends, and the wider scientific community, the future begins here.

THE SAVVY SCIENCE INSTRUCTOR

Your students may appear to be more sophisticated and mature than they actually are. Mr. H emphasizes simple, small-scale, and authentic experiences and never misses the opportunity to show his freshmen the application of science to everyday experiences. He directs his students' attention to phenomena they might never admit they don't understand:

- Boiling point? Challenge students to heat pure water in an open container to above 100°C. Many believe they can.
- Sublimation? Observe snow piles disappearing without creating a flood.
- Water pressure? Explain how a toilet works.
- Biochemistry? Find DNA oozing out of bananas and onions.
- Radioactivity? Start with smoke detectors in homes and radon in basements.

For Mr. H's students, the world just becomes curiouser and curiouser.

Connections

- American Association for the Advancement of Science
  www.aaas.org
- American Chemical Society and ACS Board–Council Committee on Chemical Safety. 2001. Chemical safety for instructors and their supervisors. Washington, DC: ACS. Available in PDF format at
  http://membership.acs.org/c/ccs/pubs/Chemical_Safety_Manual.pdf
- Flinn Scientific Catalog/Reference Manual. Flinn Scientific, Inc., 2002. Batavia, IL:
  www.flinnsci.com
- Laboratory Safety Institute. James A. Kaufman, President.
  www.labsafety.org/about.htm
- MSDS for Infectious Substances, Health Canada
  www.hc-sc.gc.ca/pphb-dgspsp/msds-ftss/index.html
- National Association of Biology Teachers
  www.nabt.org
- National Board for Professional Teaching Standards
  www.nbpts.org

(cont. on next page)

(cont. from preceding page)

- National Research Council. 1996. *National Science Education Standards.* Washington, DC: National Academy Press.
  www.nap.edu/openbook/0309053269/html/
- National Science Teachers Association
  www.nsta.org
- Vermont Safety Information Resources, Inc. (source for chemical MSDS)
  www.siri.org
  www.hazard.com

2

Communities of Learners

Promoting Science for Every Citizen

Community college doors are open wide, encouraging everyone to enter and learn. The diversity that results is both a source of strength and a formidable instructional challenge. Community colleges are charged to take students from where they enter and lead each safely to a personal best. The intellectual discipline, concepts, skills, and habits of inquiry and analysis that community colleges nurture in introductory science classes are crucial not only to those pursuing scientific or technological careers but also to all other students in their roles at work, at home, and as citizens in a democracy. Safe techniques learned in science laboratories can apply in everyday life and must be embedded in every lesson of every course.

Science for All

The National Science Education Standards (NSES) (NRC 1996) are meant for all students, regardless of learning style, background, ability, or aspirations. In the spirit of these standards, almost all high school students are encouraged to take science courses and also to pursue postsecondary education. One encouraging result is that a wider cross section of students is entering our community colleges—including some who previously might not have tried postsecondary studies. Add to this mix older adult learners, new immigrants, working adults, and myriad others, and we begin to see the wide range of students taught by community college instructors. We cannot count on a uniform level of science content knowledge, skills, and attitudes among enrollees. In many locations, we cannot even count on fluency or full comprehension of English.

As more students enroll in laboratory science classes, you, as a community college teacher, must consider a panoply of individual needs including physical accommodations, support for a variety of learning styles, compensatory support for learning disabilities, and sensitivity to language comprehension. "Science for all" challenges our capacity to provide appropriate science experiences—authentic opportunities to collect and analyze data in a safe environment.

2

Heterogeneity in classes may be just as challenging for veteran instructors as for the newest faculty member because it requires new techniques and methods. The more we find out about learning styles and intelligences, the more we realize that the old methods of sorting students by ability or future plans are not only inappropriate but also ineffective. The strategy of teaching as if every enrollee were headed for a science major simply won't work anymore—if it ever did. The wise instructor has seen changes in classroom composition as opportunities to enrich the curriculum and turns to the body of research on new and more effective instructional strategies that help make science comprehensible and safe for everyone.

Treasuring Diversity

A heterogeneous class has value to both the students in it and to society. Community colleges are preparing citizens and employees who must work together at all levels. The usefulness of skills students develop when they work in groups with fellow students of differing abilities is hard to overestimate. Sensitivity to the gifts, quirks, and needs of others has value in the workplace as well as in the community college laboratory. The cooperation and the teamwork skills students learn from communicating ideas in different ways will help them in many life circumstances.

The tone for appreciating diversity begins with the instructor. An instructor who values and respects the differing skills that students bring enriches the experience of all students. If an instructor considers modifications and adaptations onerous duties, the class will sense it and respond less fully.

Working Groups

You may have thought of arranging laboratory work groups rather than letting students choose their own partners. If not, consider the variety of skills, besides reading and writing, that are required for successful laboratory work. Students must manipulate instruments, observe carefully, record accurately, and communicate among themselves. No single student will excel in all areas, and students who lag behind their peers in one skill may be quite advanced in another.

By thoughtfully assigning lab partners and groups and monitoring and coaching these teams, you can create situations in which students appreciate each others' strengths and capitalize on differences in abilities to produce a whole greater than a team's individual parts. One way to find out about students' skills and challenges before assigning lab groups is a self-survey: Ask them, for example, which of several roles—note keeper, equipment manager, direction interpreter—they prefer. Then assign groups with at least one student in each area of strength.

SCILINKS. THE WORLD'S A CLICK AWAY

Topic: learners with disabilities
Go to: *www.scilinks.org*
Code: SSCC16

Course Materials

It is rare that a single textbook or laboratory manual can accommodate all the students in a course. Instructors must, with assistance from others in the institution, find or prepare alternative or supplemental materials and ensure that everyone understands the key concepts of the course, directions for laboratory investigations, and safety precautions.

2

Large print, audio, or native-language support may be needed. Graphics and other illustrations may assist in comprehension, but instructors should review all graphics, including signage, to ensure that everyone understands the meaning intended. Fortunately, the internet is a rich source of supplemental material. In many of this book's sections, particularly in the "Connections" at the end of each chapter, you'll find sites to which you might link to provide extra support for students.

If your institution has a course platform, which permits teacher and student interaction via the internet, consider supporting your face-to-face program with online resources. You can set up forums for questions that students are reluctant to ask in class and for specific language discussion groups, assuming you or another knowledgeable speaker of the language can monitor the discussions.

Accommodation Is the Law

Adapting programs for every learner is a serious legal and ethical obligation. It's one the science instructor may not avoid or pass on to other professionals. There is state and federal legislation that requires us to reduce or eliminate physical barriers to everyone and, in some circumstances, to educate students in "least restrictive environments," meaning to the maximum extent appropriate with their nondisabled peers. That doesn't just mean giving disabled students a chance to take classes but also creating a system that supports their success.

More and more, community college students present their instructors with requests for accommodations. Some have well-defined needs and can be quite specific in the accommodation they are requesting—a little more time or a quieter place for testing, a seat at the front of the room, probes rather than measuring instruments with fine numeration, and adaptive hearing devices. Others may be vague. They want to attempt the standard program, but they need a parachute if their efforts don't succeed. Then there may be students who are entitled to accommodations—or who can benefit greatly from them—but who do not realize that assistance may help or are too reticent to ask. In any case, check the legal obligations to accommodate and give thought to the professional responsibility instructors have to provide whatever assistance is possible.

Inclusionary obligations may seem a burden, but they carry distinct advantages. Inclusion prompts us to look at programs in new ways. We can ask ourselves what we are really trying to teach and what part of our program is there simply because of tradition. By analyzing what really counts, community college instructors can enable a wider range of students to achieve far more than anyone had ever thought possible.

> **An ADA Checklist**
>
> Check out your laboratory. Here's a sample of ADA guidelines:
>
> - 86 cm wide aisles for wheelchairs with appropriate turning radii
> - 70 cm of knee space
> - A sink no higher than 86 cm and no deeper than 17 cm with paddle handles
> - Paddle handles at sinks and on doors
> - All entrances wider than 86 cm
> - All flooring leveled or ramped
> - A clear emergency exit through accessible doorways (not automatic fire doors)
> - Clear sight lines from a sitting position
> - Locked storage in appropriate forms
> - No protruding cabinets
> - Access to the safety shower in 10 seconds
> - Braille labels on safety equipment
>
> What if your lab does not comply? You may have to redesign an activity so a disabled student can safely perform it. You may not exclude the disabled student from the activity or provide a substitute activity that does not meet the essential goals of your program.

The Americans With Disabilities Act (ADA) of 1990

The Americans with Disabilities Act (ADA) of 1990 (*www.usdoj.gov/crt/ada/adahom1.htm*) prohibits discrimination against people who have significant impairments to basic life functions. ADA requires educators to provide access to facilities and programs for all students, as well as other members of the community—instructors, students, and the general public. Any institution that receives public funds of any kind is subject to the law. (Most private schools accept some public funds.)

If an older public building—or a private school receiving federal help—is not handicapped accessible, under ADA, it may not need to be accessible until remodeled. People often (inaccurately) refer to this as "grandfathering." But even "grandfathered" institutions must create plans under ADA for steady progress toward compliance.

Section 504

Section 504 of the Rehabilitation Act of 1973 states specifically that no "otherwise qualified handicapped individual" shall be excluded from participation in a program or activity receiving federal financial assistance. This is the part of ADA under which most students qualify for and request accommodations at the college level.

A student may be academically gifted and yet have a handicap that qualifies as a disability. Physically handicapped students, even very high achievers, require accommodation under ADA. Attention-deficit/hyperactivity disorder (ADHD) is a disability affecting adults as well as children that may require accommodations under ADA even though it is not an apparent physical disorder. You may also need to be sensitive to communication disorders, such as Asperger syndrome or aphasia or to conditions that require students to take frequent breaks or intermittent medication.

Because the student must initiate the request for accommodation, many do not avail themselves of services that would help them succeed and consequently fail.

Community colleges have, or should have, a disability services director not only to substantiate disabilities (sometimes by testing) and set up necessary accommodations but also to help instructors recognize the need for accommodation and to encourage students to seek the accommodations they need. Common services provided at the community college level include tutorial centers, note takers, and extended time or other means of support during testing.

Each state has regulations governing how schools should comply with these federal mandates; some state statutes go farther than the federal mandates. Consult your institution's disabilities coordinator to make sure you have taken every step needed for appropriate accommodation.

A Good "IDEA"

The Individuals With Disabilities Education Act (IDEA, formerly Public Law 94-142, the Education for All Handicapped Children Act passed in 1975), reauthorized most recently in 1997, mandates that all students who have not completed a high school education receive a free and appropriate public education regardless of the level or severity of their disabilities.

You may wonder why, in a guide for community college instructors, we are quoting legislation that protects students who have not completed high school. In increasing numbers, students who have not received their high school diploma are enrolling in community college classes. They may be enrolled simultaneously at both levels. If these students qualify for IDEA services, services may be required in the community college setting as well as in high school.

A student qualifies for services under IDEA if he or she has a disability that interferes with learning. IDEA requires that, to the greatest extent possible, students with disabilities be educated with students who do not have disabilities. The law states that "unless a child's Individualized Education Plan (IEP) requires some other arrangement, the child is (to be) educated in the school which he or she would attend if not disabled [Section 121a.522(c)]." It permits the removal of the student from the regular classroom only when education in regular classes "with the use of supplementary aids and services cannot be achieved satisfactorily [Section 121a.550(2)]." This means that, if it is possible and practical for a student to learn a subject in a regular education classroom, it must happen that way.

Another mandate of IDEA is a vocational plan. Such a plan may specify a sequence of courses or identify a specific set of skills to be acquired. (The most common dual enrollment for a special needs student is in an applied or vocational course.) When skills are listed in a plan, there must also be an evaluation plan to track the acquisition of those skills on a continuum—separate from the more familiar grading system. You would need to familiarize yourself with the skill set identified for students that have plans and may need to make adjustments in the grading of these students.

2

Some students may have accepted their diploma, but still be following the guidance of their secondary vocational plan even though they are no longer entitled to the legal protections of IDEA. Whether by the letter or the spirit of the law, those students must also be supported in their career goals. Often, those goals include community college courses—and that means you.

ALTERNATIVES TO WHAT IS WRITTEN

No student can be excluded from a course or an essential part of it based upon disability. If your course begins with a safety test, you may need to develop a different form for the language-disabled student, or you may need to enlist the help of a consultant to verify that the student can understand and follow safety directions. You may want to enlist help to put a form of that test in Spanish or another language on your online platform.

You cannot simply exclude a learning-disabled student from an activity on the basis that he or she failed a particular safety test or quiz. You must develop some remedial plan for instructing and testing those who would fail your regular assessment. Make safety evaluation a regular event, not just a start-of-term special. If a student fails to understand a written direction, you could be liable for an accident that occurs.

Easier Said Than Done

To comply with the laws on handicap accommodation, it's essential that you understand the core goals of each course you teach. To make an informed judgment on whether a student's request for accommodation is reasonable, you must be able to separate the benchmarks of your course from what has traditionally been expected from students.

Consider these examples. A student who is visually impaired enrolls in a biology laboratory. The measuring instruments require fine visual discrimination. The student requests probes or other alternative equipment. This request is appropriate and reasonable. The same would be said of a hearing-impaired student who asked for assistive devices, or a student with reading disabilities who asked for more time on tests and books on tape.

A more difficult call might be the student with documented disabilities in mathematics who requested accommodation in introductory physics. It would be reasonable to provide computer-assisted tutorials in the learning laboratory, but eliminating the mathematics content from assessments in a course that was intrinsically based upon math probably would not be considered a reasonable accommodation.

When it comes to safety, instructors must make every effort to accommodate every disability. Having alternative versions of safety directions on tape, in large type, on your course platform, or even in short video clips on a website are all reasonable approaches to possible requests for accommodation. However, we can't expect that students know what to ask for in the way of accommodations. If the sinks are too high for a disabled student to use proper hygiene, or the hot water pipe is in the way of a paraplegic's numb knees, the instructor should anticipate the problem. If the lab directions are "add acid to water," instructors must think ahead: Are there students in this room who need more reinforcement in a different form?

Accommodation laws require the full cooperation of the college administration. The science department chair may need to update administrators on how IDEA, ADA, and other legal mandates affect the science program and facility requirements.

If a student with a physical disability enrolls in your course, you should know all the details in advance. It is the school's responsibility to ensure you have all the equipment you need. That may mean different furnishings (see "An ADA Checklist" p. X), Braille, text on tape, sound amplification equipment, earphones, personal word processors, or other assistive devices. You have the right to participate in formulating an IEP, and to request the help you need to modify your program and work space. Maintaining a good balance between order, accessibility, and open inquiry will take time and effort, but it's always easier with a team. Don't forget you will have to spend time planning and preparing those lab and support personnel (see Chapter 11, "A Diversity of Needs," p. 176). It takes special support from a department head to help adjunct instructors meet the accommodation requirements of students, especially if the sections are scheduled for weekends or evenings when daytime staff are not around. The key to making all students welcome in college science is teamwork.

Creative Thinking

Use your eyes and ears creatively to make sure your classroom doesn't present barriers to students. Some professional preparation programs require prospective instructors to spend time in a wheelchair, on crutches, or with blurred vision or muffled hearing. The experience usually gives the instructor a very different perspective. The suggestions in this chapter do not cover every possible barrier, but they can provide you with a place to start, especially important if you are in an older building or an adapted classroom.

Begin your observations by looking at the physical facilities in which you instruct. To accommodate a student with a physical disability, you will need extra space—probably twice as much as for students who are not physically impaired—and specialized equipment. A wheelchair may be as wide as 86 cm and may take up even more room if the wheels are cambered, or tilted out, for a paraplegic. Wall-mounted objects should not be higher than 86 cm from the floor, and there should be at least 70 cm of knee space under the desks. Many people with disabilities must sit on special cushions to prevent pressure sores. This increases knee space requirements. Sinks must not be more than

2

17 cm deep and must have paddle handles to accommodate people for whom turning knobs would be a problem.

The floor must be flat, including the path to the safety shower, and there should be no barriers such as taped-down wires or uneven carpet/tile interfaces. Make sure there is a good clear exit path from the room in case of fire. Don't rely on a route through a fire door that may close automatically if the fire alarm sounds.

Think about visually impaired students as you inspect your room. You may need Braille labels. Wall-mounted units should be placed above base cabinets. There should be no protruding edges or corners on casework and furnishings, an accommodation for visually impaired students that is valuable for everyone. You should also be conscious of acids, glues, or solvents that can make fingertips lose their sensitivity, a problem for students who read Braille.

Think also about students with hearing impairments, even minimal or frequency-limited disabilities. Allergies and overuse of loudspeakers can cause temporary hearing disabilities, too. Learn to distinguish between unnecessary noise and the good noise of organized bustling. Insist on a businesslike silence when you need to provide instruction.

The building design should conform to ADA requirements. If it doesn't, don't rely on a grace period if a student with a physical disability is enrolled in your course. You must create an immediate plan for changes when a student with a disability needs access. Until a major remodeling project occurs, you probably will need to add portable lab stations, adjustable-height tables, and alternative sink stations.

It's Not What You Say, It's What They Understand

Most instruction during both lectures and labs depends upon verbal communication—written and spoken words. The wise instructor does not depend on head nods and smiles to signal understanding.

Can You Hear Me Now?

Headsets, loud concerts, and the environmental noise accompanying our modern society have had a serious detrimental effect on hearing. In the good old days, hearing loss was more often associated with aging. Today, young people may suffer greater hearing losses than their grandparents.

When hearing is damaged by abuse or injury, the full range of sound waves may no longer be interpretable. It is common for hearing to fall off significantly at the higher range of frequencies with those pitches missing, attenuated, or distorted. The effect can be detrimental to the students' comprehension.

Amplification alone does not clarify speech when hearing loss is limited to specific frequencies or frequency ranges. Speaking more loudly will not guarantee that the message will arrive correctly.

We define communication as the sending of a message by one person and receiving of exactly the same message by another. For someone who has limited hearing caused by losing some of the hearing range, strange effects sometimes take place. The brain will try very hard to make sense of the nonsense that the ear is reporting, translating and presenting the individual with a message, which may be incorrect. For example, during a lecture, when you mention *moths*, a student may be nodding in agreement with apparent understanding. If you ask, "Is this clear to you?" you may receive an honest "yes." Unfortunately, the student may be reassuring you that he or she understood your point about *moss*. Try to keep your antennae up for potential misunderstandings.

Hearing difficulties are varied and often undetected or underdetected. In some cases, sound systems will help. In others, medical examination and hearing enhancing devices (hearing aids) may be needed to restore full hearing range. The person needing intervention may not be aware of any limitation. Some students may not be working up to potential due to undetected hearing difficulty and never realize that their problem is a hearing disability rather than a lack of intellectual ability.

Be particularly aware of hearing issues when safety is involved. During lectures, misinterpreting a concept may affect only a grade. Miscomprehending safety instruction may be much more serious.

Signal Versus Noise

Institutions can provide instructors with voice amplification systems to help them and their students communicate comfortably. Raising the signal in relationship to noise (S/N, or signal-to-noise ratio) can be more important than raising your voice. The signal is the sound the listener is attempting to hear or distinguish, and the noise is the ambient noise in the room. For speech to be heard clearly, the sound to be heard must be loud as compared to other noise such as side conversations, the whir of ventilation fans, and the hum of machines. This is the idea behind the use of FM (frequency modulation) amplification systems. The instructor's speech is picked up by a microphone on the instructor's lapel and delivered amplified directly to the student's earpiece so, to the student, the instructor's voice sounds louder than the ambient noise.

Limited English

Community college is often an entering point for newcomers to our country. Many immigrants, though well educated, may have limited English ability and look to your course to learn English as well as your subject. Others may have not only limited English ability but also limited formal education. English language learner (ELL) or English as a second language (ESL) students arrive in class with varying levels of comprehension. Some have little or no English skill, and some have picked up a little here and there. Others may be conversationally adept but lack familiarity with common idiomatic phrases.

2

In many cases, students new to the country may be anxious due not only to English difficulties but also to unfamiliarity with customs and expectations. They may enter your class uncertain of acceptance, fearful of rejection or doing something wrong, and consequently nod, smile, and answer "yes" when you ask if they understand. In some cultures, the smile indicates respect, not comprehension.

If your class includes students with limited English ability, you need to make sure that your procedures are clear to them. You may have students who do not follow directions because they can't read or translate them. It is important to understand each student's level of English proficiency. That may mean requesting the results of a comprehension test or interviewing the student.

Safety instruction must be tailored to each ELL student's level of comprehension. A student who is nearing proficiency in English may be able to comprehend regular classroom material. A student with very limited English and little reading ability requires a completely different approach. As students work through the stages of proficiency, they need communication designed to accommodate each level. Safety information must be communicated by the instructor and comprehended by the learner. Communication and comprehension are the issues—requirements for a safe learning environment.

Invest some time in considering how you will assess comprehension. How will you know that the issue of keeping exits clear at all times, or the shower location, or any of your safety lessons was comprehended? That nodding head and smile may have masked an anxiety-driven confused mind unable to make sense of your instructions. Assessment is as important as presentation: The best presentation is meaningless if the message does not arrive intact.

You should have safety signage that uses universally understood symbols and/or is in the native languages of your students. In some areas, Spanish/English signage should be the norm. Where there are known concentrations of other languages, such as Creole in South Florida, Asian languages in California, and lesser-known Chaldean in Michigan, ask for help from the local community in preparing safety warnings. For students who receive ELL support in languages that aren't common but occasionally occur, ask your institution's ELL instructor for help.

It's not just students who may have limited English. Colleges value—and hire—science instructors who are not native English speakers. All laboratory instructors, nevertheless, must be able communicate clearly enough in English that standard and emergency directions given by the instructor can be understood by all students in the class. It's not enough to speak English so that native English speakers can understand; instructors must also be able to communicate standard English in a way that students of Hispanic, Creole, and other linguistic backgrounds can clearly understand. The institution is responsible for ensuring that instructors as well as students have adequate and appropriate language support to communicate clearly. This safety requirement is paramount.

Different Strokes

In every college, many students have what might be called undocumented social handicaps. Some are chronically absent because of poor planning or heavy family responsibilities. Some pay the tuition only to remain technically dependent on their parents; a few are virtually homeless, moving from friend to friend. The relationship problems that can beset college students can be overwhelming. Even high achievers may have so many other responsibilities that they're frequently absent, late, or leaving class early. The side effects of what's happening in your students' world can affect the function of your classroom, causing not only distractions but also real safety problems.

With your sensitive "instructor antennae," you have to distinguish true social crises from the constant buzz of college sociology. Is that young woman in the back of the room crying because she just broke up with someone, or might she be pregnant? Is the foreign student in front quiet because he is unsure of the answer or unsure of his welcome? Science instructors often hear more than others because students are working and talking in groups during labs. It's important to know when to the tell students to get down to business and when to take time to resolve pressing issues.

Some college students have a low sense of fate control. In the words of the late science educator Mary Budd Rowe, they are "dice players"—believing that what happens to them in school occurs by chance rather than because of their own actions. Dice players think that instructors make up grades and that achievement is out of their control. Low achievers often can't plan beyond today or tomorrow. For these problems, it's important to have a tight, consistent, and transparent grading structure. Have a system that encourages students to check their own progress at frequent intervals. Grade for small, short-term achievements because many students can't guide today's behavior based upon a potential grade a week away. And always reinforce the idea that success is within a student's reach.

Your syllabus should be clear enough to define your core expectations and the consequences of failure to do the work of the class. That should include the consequences of potentially unsafe behavior. There should be resources for those who cannot comply easily or immediately, but ultimately students must share the responsibility for safety. There is no disability, no social crisis, and no conflict that should justify a student's participation in a laboratory activity in such a way that it would endanger the student, his or her peers, or the college staff.

To Everyone's Good Health

Sometimes disabilities may be temporary—and contagious. The science class might be an ideal place to make clear that infectious diseases are communicable and that reasonable people avoid infecting others. Be conscious of the possibility of diseases transmitted via blood and other body fluids. For information on Standard (Universal) Precautions, see Chapter 10, pp. 156–157.

2

> ### Fighting Infection
> - Teach basic health precautions as part of your curriculum.
> - Ask students to avoid attending classes while infectious.
> - Keep tissues handy and reinforce use and proper disposal.
> - Keep soap near the sink and encourage hand washing.
> - Do not allow students to dispense drugs to one another in class. (Many students carry a few aspirin or ibuprofen. Do not become involved in exchange of these items.)
> - Ask that your room be kept relatively cool.
> - Keep nonlatex gloves handy.

Although allergies aren't infectious, hygiene rules for coughs and colds apply to allergy symptoms because infections can take hold in allergy-inflamed tissues. Be aware of the possible presence of allergens in your room. Provide an easy and private means for students who have allergies to let you know, and make sure that things you keep around your lab do not exacerbate allergies. See Chapters 5, p. 72, and 10 "Persistent Problems," p. 153, for more information about allergens. Remember that allergic reactions can become life-threatening conditions very quickly. If a student develops hives or any sign of respiratory distress, call for medical help immediately.

Do not provide medication—prescription or over-the-counter—to any student. Even the most common over-the-counter medication can cause a severe reaction. Never administer medicine of any kind. You are not qualified, authorized, or insured to do so. Discourage the practice of students' bringing medications (except emergency medications) to class. When a student must take a medication at school, it should be transported in single doses in labeled containers.

Instructors should also learn to recognize the signs of substance abuse—actual use and the signs of long-term damage after use. The exact cause of a problem is not always clear. College students who are chronically sleepy or lethargic may be working too many hours or may be abusing drugs. Those who can't seem to sit still may have attention-deficit/hyperactivity disorder or may be coming off the effects of drug experimentation. The student services office is invaluable for helping with students who are not able to follow safety directions in your classroom.

A Little Help From Your Friends

Whether you are full time or adjunct, instructor or administrator, you should not be expected to make every accommodation decision on your own. Each institution should have a team of people trained and ready to help make the calls on what is possible and what is reasonable.

Though a large college may have its drawbacks, it may also have more resources. Counselors, disability coordinators, ELL specialists, learning laboratory staffs, and administrators can help make adapting to diversity possible and practical. Dual-enrollment students may have access to specialists in their high schools. Set a goal of having every student succeed safely every day, and you'll have a whole team of cheerleaders to help you succeed.

Every syllabus should have the name and contact number for the school's disability coordinator clearly listed. In addition, list a method through which students can easily and privately request accommodations from you. An e-mail address is probably the best private and practical way. Tell students that you are ready to respond to reasonable requests for variations, but that you can't help them after the fact and can't respond to requests that they don't make.

We as educators have the responsibility to do our very best to educate our students—every one of them. How a student arrived at our classroom door is not a consideration for instruction. We must accept all, openly and warmly, doing our utmost, preparing them to acquire a quality of life better than that they had when they arrived at our door. A number of resources that you and your institution may find helpful are listed at the end of this chapter.

THE SAVVY SCIENCE INSTRUCTOR

Ms. J's an adjunct—but most Monday evenings, she feels like an extra. She gets the extra sections and the extra students. Sometimes there are enough newsletters, copies of memos, and equipment—sometimes not.

Because Ms. J's name isn't listed in the course catalog, she sometimes has extra seats on the first night of class. But by the second meeting, she normally has a lab full of students who have dropped out of other classes. On those rare occasions that she gets to share her anecdotes with other faculty, she gets the impression that her group has the highest diversity and the most special needs. More and more students arrive with the vague requests for accommodation that students make.

By the time Ms. J gets to campus each Monday evening, the disabilities office is closed and most of the lifelines that support other instructors aren't available. So this year, Ms. J put in a little extra time before the term started to create her own resource bank for students. She constructed a website with accommodations that might help special needs students. Now, when students hand her requests for extra help, she offers them the link and asks them to consult with their advisers to select the accommodations most appropriate to their disabilities: extra testing time or alternative test locations, larger print books, audio files of safety directions, probes instead of instruments with small numbers, tutorial materials, or voice amplification software.

Now the responsibility is on the student, and Ms. J has extra time to plan for a safe and challenging course.

2

Connections

- ADA, The Americans With Disabilities Act
  www.usdoj.gov/crt/ada/adahom1.htm
  www.usdoj.gov/crt/ada/pubs/ada.txt
- American Chemical Society. 2001, 4th ed. Teaching chemistry to students with disabilities. Washington, DC: ACS.
  http://membership.acs.org/C/cwd/teachchem4.pdf
- Barrier Free Education
  http://barrier-free.arch.gatech.edu
- CEC, The Council for Exceptional Children
  www.cec.sped.org
- IDEA, Individuals With Disabilities Education Act
  www.ed.gov/offices/OSERS/Policy/IDEA/index.html
- National Research Council (NRC). 1996. *National Science Education Standards.*
  www.nap.edu/openbook/0309053269/html
- West Virginia University, Inclusion in Science Education for Students With Disabilities
  www.as.wvu.edu/~scidis
- U.S. Census Language Use and English-speaking Ability: 2000
  www.census.gov/prod/2003pubs/c2kbr-29.pdf
- National Clearinghouse for English Language Acquisition
  ww.ncela.gwu.edu
- National Science Teachers Association. 2000. Position Statement on Multicultural Education
  www.nsta.org/positionstatement&psid=21

Where Science Happens

3

Equip Your Lab for Safety

Since the building boom of the 1960s, community colleges have housed generations of instructors and learners, weathered enrollment ups and downs, and borne changes in educational practice. They are almost always near the communities they serve which, in turn, support them. Despite continued high enrollments, many community colleges have not had the modernization and expansion needed to support best practice. Even when there is plenty of space for laboratory science classes, the facilities may not support investigative programs safely.

The Practice of Science

Whether students intend to pursue careers as laboratory workers and technicians, prepare for four-year college studies in the sciences, or fulfill general studies course requirements, it is essential that their science courses include laboratory experience. They must learn science by observing phenomena, isolating and manipulating variables, gathering data, and critiquing and analyzing results. For this they need physical facilities—space, furnishings, and equipment—that support safe laboratory work. The best instructor cannot operate safely without the right physical space and equipment. In the wrong facilities, science investigation poses serious safety hazards.

Older science facilities may have been built for different student populations or completely different purposes. If they have not been retrofitted or renovated recently, they may lack technology support and modern safety equipment. Even new facilities can be too crowded or architecturally unsuited for safe laboratory work. Science labs that were designed specifically for chemistry, biology, physics, or Earth and space science may be inappropriate if flexibility was not built into the design. The best new labs are built with the assumption that change will occur and are designed to safely accommodate laboratory work in any science discipline with minimal modification.

3

Take Out Your Tape Measure

Although the following standards were originally developed for high school laboratories by organizations such as the National Science Teachers Association (NSTA) and architectural and fire protection groups, we recommend them as minimum standards for community college laboratories:

- a minimum of 4 m^2 per student (96 m^2 for a class of 24) for a laboratory, 5 m^2 for a laboratory/lecture room
- additional and specialized space to accommodate students with disabilities
- 1.4 m^2 for each desktop computer station
- 0.9 m^2 per student of preparation space for the instructor
- 1 m^2 of lockable storage area for every student in the room
- a ceiling height of 3 m
- two escape routes (a second door or large window without screen)
- ventilation of at least five (eight changes are now recommended) air changes per hour, with fume hoods for every chemistry or biology laboratory group
- hot and cold running water with soap
- safety eyewash and shower facilities with tepid water
- fire protection and fire safety equipment

Source: Biehle, J., L. Motz, and S. West. 1999. *NSTA Guide to School Science Facilities.* Arlington, VA: National Science Teachers Association.

To facilitate communications and to permit lab instructors to adequately monitor work being conducted during laboratory activities, we strongly recommend doing away with laboratories designed with long rows of double-sided lab benches obstructed by eye-level shelving and separated by narrow aisles, or L-shaped laboratories with limited sight lines and barriers to verbal and visual communications.

Give Me Space, Lots of Space

The single most important safety factor is space. Sufficient space in the laboratory is critical to a safer program. Research that relates the amount of workspace per student to accident frequency supports the conclusion that when space per student goes down, accident rates go up, according to Sandra West Moody, Southwest Texas State University. With enough space, a small spill is just a cleanup issue. When students are crowded, the same spill can cause multiple injuries. Elbows knock over setups, electrical cords are pulled from receptacles, equipment falls, and tensions mount. Check the "Take Out Your Tape Measure" box for minimal space requirements recommended by NSTA. Other professional associations may have slightly different recommendations, but most are similar.

A related factor is section size. Ideally, laboratory sections should have no more than 24 students even if the space might accommodate more. Research data show that accident rates rise dramatically as section enrollments exceed 24. If classrooms are small, enrollment should be limited even further. The addition of a laboratory assistant may not help—and may actually hurt—an overenrollment or space problem if the assistant takes up space but is not well trained. When either workspace per student goes down or class size goes up, the accident rate goes up. These are findings important to consider when setting caps on section enrollments.

If you cannot expand space or limit enrollment to the recommended standards, then you must *modify or eliminate those activities that cannot be performed safely in the conditions you have*. Although notifying an administrator that an activity could be dangerous provides written proof that you know a situation is unsafe, you cannot transfer your liability by notifying an administrator, nor can you delegate your responsibility to a laboratory assistant. The responsibility to avoid the unsafe situation is yours—as is the liability if you do not.

A Room of One's Own

Ideally, each instructor, including adjuncts, should have his or her own space for lab preparations and some lockable and private area for storage of equipment and supplies before and after classes. This is particularly important if the teaching spaces must be shared and part-timers do not have access to the laboratory space other than during classtime. If a science instructor sets up labs and then has to leave them, the equipment and materials in the lab may be unsupervised or subject to tampering. An instructor dispensing equipment and materials in a rush is likely to make mistakes.

Even if there is a lab assistant, there should be sufficient secure space to set up ahead of time to allow instructors to check the setup before lab work begins. Space also needs to be provided so that students who must make up work have an appropriate facility in which to work.

Laboratory work conducted in a nonlab classroom is likely to be unsafe and provide poor education. If a science course must be scheduled in a room not originally designed for laboratory work, then lockable storage, hot and cold running water, and safety equipment are an absolute minimum that must be provided. The pressure placed on a junior faculty member, adjunct, or anyone else to be a team player and take the extra section in a nonlab facility introduces unwarranted and unacceptable risk. Instructors placed in such a position should resist vigorously. Balancing laboratory equipment on tablet-arm chairs or dissecting in facilities with no hot water for cleanup is not about being a cooperative team player; it is about setting up conditions for accident, injury, and liability.

If a colleague allows a student into the chemical storeroom and that student removes the stock bottle of some reagent you ordered, could you be liable for harm that may result? It would be better not to have to find out.

3

THE ART OF PEACEFUL COEXISTENCE

Sharing classrooms is never easy. It's especially hard when several instructors with different safety or housekeeping standards must share the same rooms. Involve the department chairperson or designated administrator in establishing and enforcing rules for shared spaces, including shared storage and preparation areas. Consider getting the entire department to reach agreement on the following points:

- Recognize that all who share a room or space may share liability if something goes wrong.
- Establish and write a clear set of ground rules for use of the shared space that must be followed by all who use the space. For example:
    - All entry doors and lockable storage must be locked upon departure.
    - No students may be unsupervised in any laboratory.
    - All work surfaces must be cleaned and dried before departure.
    - Storage of ongoing projects may only be in designated areas and may not include any fragile or hazardous materials.
    - Storage of any chemicals must be accompanied by material safety data sheets (MSDSs) filed in notebooks located in the department office and in the storage room.
- Designate a bulletin area for posting written notes to all laboratory users about any unusual conditions.
- Establish a common set of standards for training laboratory assistants and assign specific responsibility for the training and supervision of each assistant.
- Prohibit students, except specifically trained laboratory assistants, from entering science stockrooms, backrooms, or prep areas at any time for any reason.
- Ensure that all unsupervised science areas are locked and secure at all times and unavailable to loitering or cut-through traffic.
- Provide professional development safety training for all staff working in laboratory areas each year. Include updates and refreshers on lab standards and Standard Precautions (See pp. 156–157).

(cont. on next page)

(cont. from preceding page)

- Keep MSDS sheets for *all* chemicals, a set of the rules, and a copy of this book in a location regularly accessed by all staff who use the shared space (labs and prep areas).
- Ensure that stockrooms and chemicals are secured after the teaching day when other classes or community groups may use the classroom.

A Lab for All Sciences

Should a classroom for biology be designed and built differently than a classroom that supports chemistry? For contemporary science curricula, the best answer is no. In college instruction as well as in research laboratories, the fundamental elements of a safe facility are the same. Although physical science equipment may be longer, wider, and heavier than equipment for other sciences, and biology instructors may need microscopes while chemistry instructors have more use for spectrometers, all instructors need emergency equipment, eyewashes, showers, protective eyewear, clear and level counter space, secure storage, prep facilities, direct outside air ventilation, hot and cold running water, adequate utility services, and good lighting.

Course selections will vary from year to year, curricula and degree requirements will certainly continue to change, and the separation between science disciplines will become less distinct. In the long run, the most effective way to accommodate a full investigative science program is to equip all science rooms for safe laboratory work and vary the furnishings to suit the discipline.

Floor Plans and Floors

A floor plan that distributes workstations as far apart as possible on perimeter walls has several advantages. Properly designed, each standing-height workstation with sink, water, gas, electric, and telecommunications service can support two lab groups working at countertops on each side of the services, so six such stations can accommodate 24 students working in pairs. Perimeter workstations maximize separation between lab groups and position students facing away from one another, which minimizes distraction and interactions between groups. Accidental sprays or spills are more likely to hit walls or windows rather than another working group. If you choose additional sturdy but moveable tables that are the same height as perimeter counters, you can extend work areas into the middle of the room when needed but maintain clear space and flexibility to accommodate activities from many different science disciplines. The tables can also be rearranged for discussion or lecture.

3

Older-style floor plans with large, fixed lab benches taking up most of the floor space in the room are much less flexible and do not lend themselves easily to class discussions and large-group instruction, necessitating separate lecture areas. In some cases, the lab benches create work areas that do not have clear, straight paths to exits and safety equipment. Students, especially those with mobility problems, could be trapped or seriously slowed down in case of fire or accident. If workstations must be placed in the middle of the room, try to have compact service islands incorporating sink, water, gas, electrical, and telecommunications service, and use moveable tables that can be pulled up to the service islands in a variety of configurations. Eliminate the old-style, high shelving on lab stations that obstruct visibility and communication.

Flooring can be a safety issue. Carpeting should not be used for science rooms (see Chapter 10, pp. 153–154). In older buildings, wood flooring may still be in place. This should be replaced in any renovation, making sure hazardous material—such as elemental mercury from broken thermometers—that may have fallen between floor boards or become embedded in the tiles and remained in subflooring is identified and removed. Asbestos tiles should be removed or abated (covered) by a certified expert. Seamless chemical-resistant cushioned flooring works best. Sometimes this is installed in sheets; sometimes it is poured over a prepared surface. Two other factors should be considered. Flooring applied directly over a concrete base without some cushioning can cause fatigue and impact-related foot and leg problems. Flooring that is too shiny and smooth can be dangerously slippery when wet.

Whatever floor plan and flooring you use, be sure that at least one workstation is wheelchair-accessible and that floor surfaces do not have high thresholds, or other raised structures that are impediments to safe wheelchair maneuvering, especially in emergency situations.

In the Room Where You Live

Take a few minutes to look around the laboratory with the objective eye of a visiting safety expert seeing it for the first time or conducting an emergency drill:

- Are escape routes or routes to safety showers and eyewash stations free and clear?
- Have you established an outside meeting place for everyone to gather following an emergency escape?
- If a splash should occur at any of the work areas, is there a quick clear path to a sink, safety shower, or eyewash?
- Can students get to the sink, shower, or eyewash within 10 seconds?
- In case of an accidental spill, can students back out of the way of the splash or clear the area?

- In case of fire, is there enough room to stop, drop, and roll?
- In an emergency, can everyone, including mobility-disabled persons, get out quickly?
- Can you visually supervise every area where students may be working?
- Do you leave anything unsecured on counters and lab benches after your classes?
- Are spill cleanup kits readily available rather than on some back shelf in some back room?
- Do you have specially designated sharps containers? Where do broken glass and other sharp objects for discard go?

3

Make sure your escape routes are freely and easily accessible—not just unblocked, but realistic and immediate ways to safety for every person who may be in the room.

Instruct the class members about how to help injured people. In most cases, injured people should not be moved unless they are in imminent danger of further harm. If clothing, or hair, or other bodily parts are on fire, bring the rescue equipment to the person rather than the person to the equipment.

Consider rearranging or eliminating furniture. Reduce and eliminate clutter. Do you have stores of junk? Instructors can be pack rats. Even the most spacious laboratories and prep areas can be made unsafe by clutter you, your colleagues, or your predecessors have accumulated over the years. Scan your important lecture notes and store them on a CD-ROM or flash drive. Get rid of the paper.

Science is not only a hands-on activity but must be minds on as well. Science rooms and laboratories are no place for daydreaming or careless behavior.

Packing Relief

Having to pack up and move to another workspace may be the opportunity of a lifetime. Be ruthless. Do not pack anything you can do without. Do not seal any cartons that contain items you have not subjected to hard scrutiny. Less is better. If you don't have time to look it over now, you probably never will. Anything you have not used in the past two years will probably never be used.

Water, Water, Everywhere

Every science laboratory and workroom needs potable hot and cold running water. Hot water is needed for sanitation, hygiene, and general cleanup. Hand washing is required for the vast majority of science activities. In addition to being needed for laboratory investigations, tepid water is required for the eyewash and the safety shower, meaning both must be plumbed with hot and cold water.

Six sink stations installed around the room work well for 24 students. The sinks should be large enough to accommodate buckets and other large items and deep enough so that chemicals don't splash. At least one sink in each room should

3

be wheelchair accessible and have paddle-handled faucets that can be operated with a closed fist.

Faucets can be a real maintenance headache. Well-designed faucets are solid, one piece, and are secured directly into the counter. Avoid the more common and inexpensive faucets with a separate gooseneck that can be unscrewed easily—and therefore wiggled loose or taken apart. Building codes may require that faucets prevent accidental backflow from the laboratory station into the public water supply. If you need jet-spray nozzles, make sure that they are removable and that you remove them when they are not absolutely necessary.

In many states and municipalities, water coming from science labs must pass through acid traps or neutralizing tanks before entering the regular wastewater system. Even if they are not required by code or regulation, you should consider specifying corrosion-resistant pipes and acid-neutralization tanks. If you have acid-neutralization tanks, remember that the marble chips in the tanks must be changed regularly. The frequency of this change is dependent on the volume and concentration of acids that run through the tanks, but, in general, check and recharge them at least once a semester.

EYEWASHES

You need a plumbed eyewash that anyone working in the laboratory can reach within 10 seconds. The eyewash should be no more than a meter above the floor and accessible to people who use wheelchairs. The eyewash should have a flow rate of 0.4 gallons of tepid water per minute, turn on with a single motion, and operate hands free. It must be capable of washing both eyes simultaneously.

We strongly urge that bottled eyewashes not be used, because they do not provide enough water and they can become a reservoir for infectious microbes. All eyewashes should be flushed for two to three minutes every week. This procedure helps prevent contamination by *Acanthamoeba*, a protist that lives in water and can cause serious eye infections.

In case of accident, the injured party is likely to need assistance in reaching and using the eyewash. Everyone should be trained to call for help and to provide assistance. The eyewash should be used to flush the eyes continuously with tepid water for at least 15 minutes. Students should be trained to lift their eyelids away from the surface of the eyeball when flushing.

EMERGENCY SHOWERS

Every science room needs an emergency shower that anyone in the lab can reach within 10 seconds. The shower should be 2 to 2.5 m above the floor, but the handle should be within reach of wheelchair-bound users. The spray pattern should be at least 0.5 m in diameter and cover both the affected individual and a helper. The emergency shower should deliver a minimum of 20 gallons of tepid water per minute. Since contaminated clothing must be removed immediately while in the shower, a modesty curtain would be an advantage. But if one is not built in, a fire blanket placed close by could be used.

Many safety showers were installed as part of a retrofit with no thought given to actual use or regular testing. A shower may be present but deliberately deactivated to prevent accidental flooding. To be certain the shower is ready and useable in an emergency, it should be checked and activated regularly—ideally once a month or at least once a semester. Frequent testing requires a drain and floor configuration that keeps shower-testing water from spreading. Older showers may need modification for adequate drainage, water flow, and temperature.

3

A Breath of Fresh Air

Public buildings require a constant exchange of fresh air (from 5% to 15% per hour, depending upon your state's regulations). Circulation should bring air from the floor to the ceiling and then vent it to the outside. Although the fresh air exchange may increase heating and cooling bills, it minimizes communicable disease transfer, the growth of molds, and the accumulation of allergens in buildings. Filters and ducts on air-handling systems should be cleaned according to the manufacturer's instructions.

In addition, science laboratories need a separate air exchange system that changes the total volume of air in the room at least eight times each hour for an occupied lab and four times for an unoccupied lab. Air must exhaust directly to the outside rather than through the ductwork for the rest of the building or into the space above hung ceilings. Because of the likelihood of chemical fumes being emitted from materials in use and in storage, the ventilation of science rooms and storage areas should *not* be controlled by energy-efficiency or timer systems that halt circulation when buildings or rooms are unoccupied. Be aware of the air quality and exchange in your classroom. Maintenance personnel unfamiliar with science safety requirements may alter ventilation equipment and flow to satisfy other considerations without your knowledge. Periodically test and verify that the ventilation system is operating properly. You can get great information on indoor air from the Environmental Protection Agency IAQ (indoor air quality) at 800-438-4318.

3

Fume hoods belong in *all* science laboratories, not just chemistry labs. Chemicals used for chromatography, electrophoresis, mineral tests, and geologic and biologic specimen preservation all generate toxic fumes that belong in fume hoods. This protective equipment also belongs in instructor preparation areas. If it is not possible to have adequate fume hoods in every science room, the curriculum must be modified or an exchange of rooms arranged for certain lab activities. Likewise, if fume hoods are not available for every working lab group simultaneously, then the activity should be redesigned so that students participating in the fume-generating portion of the activity are conducting it in the fume hood while other students are working on other parts of the activity. Be sure that wheelchair users have access to a fume hood.

When do you need a fume hood? If the label or materials safety data sheet (MSDS) for a chemical indicates it is toxic or a respiratory or mucous-membrane irritant, then it should be used in a hood. Use a hood for work with materials that have a high vapor pressure. In general, if the activity generates a smell, it probably requires ventilation. If regular ventilation doesn't make the smell disappear, you need a fume hood. Make sure your students know how to place their reagents and adjust the sash so the airflow is effective. Choose lab experiences that generate the fewest fumes, and consider microscale work to reduce the volume of fumes produced. Consider less toxic alternatives to fume-producing toxins. And remember, the nose is not a good indicator of toxic vapors. Some chemicals, such as formalin, are above toxic levels before you can even smell them.

FUME HOOD FLOW

Two kinds of fume hoods are commonly found in college laboratories—conventional appliances in which the velocity of air varies with the opening of the sliding sash and bypass hoods that redistribute air to achieve a more constant air velocity.

Many instructors mistakenly believe the stronger the current in a hood, the better. That may not necessarily be so. Most hoods should have an average face velocity (the average velocity of the air drawn through the face of the hood) of about 100 linear feet per minute. Too high a flow can cause a turbulence that can bring the vapors back into the room. Typically, the user stands at the face of the hood to work and places the reagent about 15 to 20 cm into the hood. The pull of the fans around the user's body creates eddy currents that can be strong enough to disrupt the experiment. Check the manufacturer's specifications on your hoods.

(cont. on next page)

(cont. from preceding page)

Fume hoods should be checked before each use to ensure they are in proper working order. Also check the following:

- An adequate supply of outside air must be available for the hood to vent properly. Venting should go directly and completely to the outside by having the exhaust stack extend well above the roofline and having an exhaust velocity of 3,000 linear feet per minute.
- There should be less than 20% variation in the face velocity (for bypass hoods only). The hood should respond to changes within 3 seconds.
- Hoods should be made of corrosion-resistant materials.
- Drain traps for water service in the hood should be flushed once a week by running water through them.

Proper use of fume hoods includes

- closing windows and doors during hood operation to prevent fumes from being drawn out of the hood.
- keeping the sash down completely except when inserting, manipulating, or removing materials from the hood.
- conducting work at least 15 to 20 cm inside the hood.
- not bypassing the fume hood or pulling reagents out for a closer look.

Seem complex? It really isn't. Some of the exercises that have been traditionally done in fume hoods probably shouldn't be done at all in a lab, while many exercises that instructors consider relatively harmless—such as chlorophyll chromatography—require fume hoods because of the solvents used.

Power Up

When instructors are asked to identify the most pressing deficiencies in their classrooms, space is often the first concern, but electric service ranks a close second. As we add more electronic equipment, there seems never to be enough electrical receptacles in the right places.

Your room should have at least four separate circuits, each protected by a ground fault circuit interrupter (GFCI). These automatic devices are not a substitute for a manual shutoff. You need both. The panic button allows you to shut off power if you have an emergency such as a student caught in a piece of mechanical equipment. The GFCI protects in cases of a short circuit to ground.

Turn on the standard equipment you would use for basic laboratory exercises.

3

If you are tripping circuit breakers, you don't have enough amperage to handle your needs. This often happens when you are drawing power for heating or cooling devices, such as hot plates and air conditioners. You cannot just add receptacles—you must add circuits to solve the power problem. Do not use the socket multipliers that are sold in hardware stores, because they will only increase the load on existing circuits.

There is no single ideal method of getting power to all workstations in a college science lab. But there are certainly wrong methods. Extension cords pose both fire and tripping hazards and should not be used. Loose cords should never cross aisles and pathways. Loose cords should never be in the way of anyone in the room. If workstations are located on the perimeter of the classroom, a power strip can be mounted above the counters for electric and telecommunications service, but this doesn't increase the amount of power available.

For electric service to the middle of the room, flexibility is important. Today's perfect furniture configuration can turn into a significant instructional obstacle tomorrow. Floor outlets that project from the floor (which architects call "tombstones") pose tripping hazards and also limit the flexibility of floor plans. Spring-loaded rollers that allow cords to be pulled from the ceiling are sometimes good options, but they are tempting toys for some students and can pull lightweight equipment off the work surface, especially if the setup is jostled or shaken. Another solution is floor receptacles that are flush with the floor and watertight when not in use. Consult an architect or your college facilities supervisor for help getting the power you need—don't try to fix the situation yourself.

Telecommunications connections should be available and conveniently located close to electrical service. Consideration should be given to where different types of electronic equipment are to be located and how to make such equipment convenient to use but also protected from spills and other damaging accidents. In some cases, wireless communication has been less expensive and safer than hard-wired connections.

My Kingdom for the Right Table

A well-designed laboratory should allow a smooth flow from seat work to lab work and back again. If you have the opportunity to select your furnishings, consider flexibility and safety together—they are compatible goals. Your furniture configuration can greatly influence how and how well instruction is delivered. Sturdy flat surface worktables for two-student teams are preferable to permanently installed casework.

At the college level, specify science-style tables with chemical-resistant tops and legs that are firmly bolted to the tabletop. Plastic laminates are not nearly as durable as heavy resin tops. Book compartments built under student tabletops may provide convenient storage, but they also tend to accumulate trash and debris. Make sure the height of the tables matches the height of countertops where sinks are installed and that there is a station accessible to students in wheelchairs.

Tables should be light enough to be pulled up to perimeter workstations or rearranged in the middle of the room for different activities but heavy enough to remain steady and immovable during lab work. If you are ordering tall—standing height—tables, select a style with sturdy legs and reinforcing stretcher bars near the base to stabilize the legs. Tables designed to be stable when in use but movable for rearrangement are best. Loose table legs are a serious hazard, so you should make it a habit to check all the tables in your room periodically and certainly after tables have been moved. Adjustable-height legs on tables and chairs are rarely adjusted and simply add to cost while reducing stability, so selecting the correct height and skipping the adjustability are the better options.

We suggest that you avoid using stools during laboratory work, because lab activities are generally safer when students are required to stand. If you have stools with backs, remove the backs to make it more difficult for students to tip back and balance on two legs or hang clothing or bags over the back.

Other Furnishings and Fixtures

As a general rule, storage for supplies should be in fixed casework and not on movable equipment. But if you must have movable storage equipment, make sure that it is sturdy, stable, and not top-heavy when fully loaded. Wheels should lock in such a manner that the equipment will not move if accidentally pushed or bumped. Tall storage cabinets not built in should be fixed firmly to the walls to avoid accidental tipping. The next chapter (Chapter 4, "Finders Keepers") is devoted to the subject of storage.

Appliances such as a dishwasher, an electric stovetop, a garbage disposal, and a refrigerator can also come in handy for laboratories or prep rooms, provided there is sufficient space and appropriate security for them. If these appliances are present in work, storage, or preparation areas, they should *not* be used for preparing or storing food or cleaning utensils used for eating. If the refrigerator is used to store volatile and flammable reagents, make sure it is spark proof and specifically designed for laboratory use.

Special equipment and other aids may be needed to ensure a safe environment for students and staff with disabilities. These are required under ADA and other accessibility legislation. See Chapter 2, "Communities of Learners," for more specifics.

Lights!

Lighting is a factor many instructors forget. The amount and type of lighting in your room can affect concentration and safety. Older students may require better lighting,

3

Fluorescent Lighting

Some fluorescent lights exacerbate hyperactivity and headaches and make computer screens more difficult to read. Some eye conditions are extremely uncomfortable when fluorescent lighting is used—vision is blurred by a "brightness" that makes labels and print difficult or impossible to read. Because the fluorescent bulbs used in most buildings are not full spectrum, true color is not seen. Full-spectrum halogen bulbs provide lighting that is closer to outdoor or natural light and may be better for people who have difficulty with ordinary fluorescent lighting.

especially when working with instruments with fine markings. Your classroom should have a light level of at least 50 foot-candles. A light level of 75 foot-candles is preferred for laboratory work areas.

Natural light is preferable to fluorescent lighting if it can be brought in without causing glare or heating and cooling problems. Indirect lighting is better for computer screens, microscopes, and other close work. If there are cabinets above work areas, you may need lighting under them. Make sure storage areas, such as the back part of deep cabinets and corners, are also well lit.

Total room darkening is needed for many laboratory activities, such as optics labs, black light usage, and microprojection, so blackout shades should be specified for all science rooms and labs. The ability to darken the area around a projector screen but keep lit the areas where students are taking notes is a definite advantage for science classrooms.

Lighting is also an important element in planning for emergencies. Investigate whether all escape routes would be visible if the power fails. Building codes usually specify emergency backup lighting for halls and stairwells, but are there parts of your lab, prep area, or storage rooms that also require this type of lighting? Keep flashlights handy, and check batteries periodically.

Hot Stuff

Many science investigations require heat in various forms. Gas service for burners is required in most of the science disciplines but not all of the time. Labs should have a main shutoff, accessible only by the instructor. We strongly recommend that the gas supply be shut off when gas is not needed.

Hot plates are often much safer sources of heat for experiments than open-flame burners. If you use hot plates, make sure they are specifically designed for lab use and are not recycled household hot plates. Check that the total amperage of all the hot plates in use at one time does not exceed the capacity of your circuits. Most hot plates have a bimetallic switch that sparks each time it makes or breaks contact. It can be a source of hazardous ignition.

Alcohol is the cause of many laboratory accidents and fires. One of our most commonly used chemicals, it is often handled much too casually. Alcohol burners may have been present in college labs for generations, but they pose multiple unnecessary hazards and their use has resulted in some of the worst lab accidents. Unnoticed cracks can cause a burner to explode, improper filling can result in flashbacks and explosions, and alcohol supply cans are a major hazard both in use and in storage. We recommend you do not use alcohol burners under any circumstances (they are illegal in some states). Alcohol and other flammable liquid supply containers should never be present in a room where burners or other open flames are in use.

Hot Glass

Glassware when heated tends to stays hot for quite a while. Many burns are caused by picking up recently prepared tubing or other glass pieces that do not appear to be hot. Be aware that you cannot tell the difference between hot and cold glass just by looking at it.

College labs may have specialized heat sources such as drying ovens and sterilizers. We do not recommend using pressure cookers for sterilization or for any other lab application. Autoclaves designed specifically for bacteriology should be located in secure preparatory rooms and used only by qualified staff, never by students at the introductory level.

Fire Protection

Some of the most serious emergencies involve fires. To meet codes, buildings are required to have features for fire prevention and fire protection. But even the best design can be defeated by thoughtless use. The keys to avoiding tragedy are prevention and preparation. It is important that instructors recognize the safety features that are part of their physical facilities and ensure that access is not blocked or obstructed.

Laboratories need two exits accessible at all times. Your room was probably designed to have two emergency exits. One may be a large window without a screen. Have you screened that window? Have you blocked a fire exit with furniture? The wall coverings and ceilings of your classroom should be fire resistant. Have you or someone else covered them with paper or hung combustible materials from the walls or ceilings? Make sure hanging projects and displays conform to fire codes for public access facilities.

You need at least one ABC fire extinguisher for your lab. More may be necessary depending on building codes and the size and layout of your facility. If you use metals (such as magnesium or sodium) you also need a type D extinguisher. Each fire extinguisher should have a nylon or wire seal tag. Visually inspect the extinguisher monthly, and notify your facilities manager if the tag or seal is missing, the pressure gauge is in the yellow or red zone, or the annual inspection is not up-to-date. Extinguishers must be recharged and resealed after each and every use.

3

Fire Extinguishers

- Class A fire extinguishers are used on fires that involve ordinary combustible materials such as cloth, wood, paper, rubber, and many plastics.
- Class B fire extinguishers are for fires that involve flammable and combustible liquids such as gasoline, alcohol, diesel oil, oil-based paints, lacquers, and flammable gases.
- Class C fire extinguishers are for use on fires that involve energized electrical equipment.
- Class D fire extinguishers are specialized fire extinguishers and may be specific to certain types of metals such as Mg, Na, and Ti.

Class K fire extinguishers are used for fires that involve vegetable oils, animal oils, and fats in cooking appliances such as found in commercial kitchens.

Water can extinguish only class A fires. Using water on a class B fire can make the fire worse, and carbon dioxide is a fuel for class D fires, so be sure you distinguish between equipment types.

For more information, go to *www.nyc.gov/html/fdny/html/safety/extinguisher/classes.shtml.*

Show every student where the fire extinguishers are located and make sure everyone knows how to use them. Many instructors use the mnemonic PASS to help students remember: **P**ull (the pin), **A**im (the nozzle), **S**queeze, and **S**weep (from outside to inside so the fire is not spread). However, be sure everyone understands that unless a fire is easily and immediately extinguished, the primary response must be to sound an alarm and evacuate the building.

Make sure smoke detectors and fire alarm signals are not obscured or blocked from access. Teach students to recognize the visual and audio fire alarm signals and respond to those signals immediately. During laboratory work, this means immediately stopping work, turning off equipment in the immediate work area, and evacuating. Post the fire escape route, and make sure students can tell the difference between a fire alarm and the signal and procedures for windstorms or earthquakes.

Before any lab exercise that involves flame or heat, remind students how to respond if hair or clothing catches fire. Make sure they know not to run or do anything else that will fan the flames, and have them practice "stop, drop, and roll." (Bring the emergency fire equipment to the victim, not vice versa.) If you have a fire blanket or safety shower, teach proper use and review the use of safety equipment before every instance in which this equipment might be needed. Science rooms should also have sprinkler systems that operate automatically.

Many of the materials you use in the classroom will have the National Fire Protection Agency (NFPA) diamond identifying the type, potential severity, and relative fire hazard that the material might generate. NFPA information can be found at *www.nfpa.org*. Remember, however, that the NFPA labeling system is 50 years old and is not intended to provide information about chronic exposure. The labeling can also underes-

SCILINKS
THE WORLD'S A CLICK AWAY
Topic: fire extinguishers
Go to: *www.scilinks.org*
Code: SSCC44

timate the potential flammability of the material in a larger fire. It was developed to protect firefighters from the hazards of chemicals under firefighting conditions. But the diamond system does provide some basic information to first responders when they confront unfamiliar material. It's also an easy graphic to teach, even to students with limited English proficiency, along with the standard icons for toxins, corrosives, and other hazards. (The material safety data sheets discussed in Chapter 4, pp. 56–59, provide more accurate information on chemicals you use or store.) Local firefighters or your state fire marshal may be willing to do a courtesy inspection to advise you of ways to make your facilities and lab practices safer.

Signs and Symbols

Signage and labeling in your laboratory enhances safety. Make sure the room number is clearly displayed outside and inside your room and that your students know what the room number and location are. Post the room number, location, and important emergency phone numbers and extensions at the telephone or intercom and right beside the fire, storm, and earthquake procedures. Indicate where supplies and special safety equipment such as the eyewash or fire blanket are stored. In an emergency, you may be incapacitated, so students need to know how to respond and how to give clear directions to emergency personnel.

Make signage an integral part of your room organization plan. Label where things belong and where they do not go—"no paper products in this cabinet," for instance. Use graphics and symbols as well as worded labels and instructions. Use multilingual and Braille signage as needed. Display instructions for maintenance and organism care responsibilities. Provide prominent indication of the locations of all emergency equipment, such as eyewash, shower, and fire blanket, and keep the equipment accessible. Keep eye protection on display with signs describing proper use and cleaning.

Clearly written and readable reagent labels are vital to safety. All chemicals, individually dispensed portions as well as stock containers, need accurate labeling. There should be no unlabeled reagents in your classroom. Labeling is not just for you and your students. It is also essential for the safety of the support staff that clean and service your facilities. For more on labeling, see Chapter 4.

Beyond Bricks and Mortar

Facilities are the least flexible part of a college program. Although there may be long-term remodeling goals for most buildings, many instructors find their current situation is very limited. What do you do if you know your classroom doesn't meet safety and facility standards and there is no immediate relief in sight?

First, clean up and clean out. Make space by removing everything you don't need.

3

You may need to box up and label materials by course title and topic and store them elsewhere. Then take a hard look at your furnishings. Can you invest in more flexible pieces or trade with another instructor? Can you rearrange to create more space?

Prioritize your maintenance requests, and document them. Don't fall victim to the they-won't-do-it-anyway attitude. Repeat your requests at regular intervals, and explain to your administrator what choices you are forced to make until you can get repairs made.

Meanwhile, downsize your lab experiences and your storage needs. Then remember that science is everywhere. While you wait for better facilities, use the facilities you have to expand your students' understanding of the world—and of safety. Let the administration know all the positive things you have done and what you cannot do because of your concern for safety and liability.

WARNING

You must document safety problems and see to it that this information is communicated to the appropriate authorities, but this does not relieve you of responsibility. If you know an activity is unsafe and still perform it with your students, you can be held liable for purposefully and knowingly placing students in danger. Documenting a safety problem in writing is strong evidence that you know an activity is dangerous. You do not want to hear an attorney say, "So you continued to do a dangerous lab even though you knew the facility was not equipped to do it safely?" See Chapter 11, pp. 177–179, for more information concerning liability.

Building for the Future

Be careful what you ask for—you just might get it. A renovation project may bring safer lab activities within your reach, but the process of renovation has potential hazards and inconveniences built in. If you have contractors on-site for renovations, you have special challenges for your program and for the safety of your students. Be conscious of ventilation patterns that might be disrupted. If your room will be temporarily less well ventilated, or if you will have less natural light, modify your program accordingly.

The renovation project should have an on-site supervisor who is familiar with the site and can act on behalf of everyone who must continue to work in the building. Make sure the responsible contact person is aware of special considerations for maintaining safety for your laboratory program. Make the construction contact person aware of any potentially hazardous materials or equipment present in your classroom—especially those things that could be harmful to construction workers if there were a spill or other emergency. On-site supervisors know their job, but they don't necessarily know yours.

In preparation for the renovation, review your plans and consider simplifying and revising some of the activities. You are likely to find you will not be able to accomplish as much in your science program. You and your students may be more distracted, exhausted, or both as the construction project extends beyond a few weeks or a couple of months. On days when excessive noise or other disruptions are predicted, consider taking classes outdoors for fieldwork or off-site on a scheduled field trip. Above all, be flexible and maintain a positive outlook.

THE SAVVY SCIENCE INSTRUCTOR

The best thing about Home Town College wasn't its ivied walls. In fact, there wasn't any ivy—and there were hardly any walls. The college was mostly virtual: Its introductory programs were scheduled in rented facilities and the subsequent classes were online.

Ms. F loved the college's philosophy and its dedication to students. She loved the curriculum that she was to teach the introductory students during those first vital weeks of their program. But oh, those facilities! The labs that HTC rented were just plain unsafe. The chemicals were stored haphazardly, the fire equipment had obscure tags with dates reminiscent of Elvis, and the refrigerator was positively filthy.

Her first response was panic. How do you teach safely? Even more importantly, how do you teach laboratory safety when this host school's example is so poor?

Two deep breaths, then she made a plan. Ms. F juggled her first few lectures to relocate them outside. Then she brought in the dean and the head facilities person. No safety—no occupants—no rent. It was an ultimatum, but an effective one. By the time the classes returned to the lab, the suspect chemicals had disappeared and the furniture was clean. The following day, signs of a bonded disposal company were seen on the grounds. By the next week, even the refrigerator looked as if it had been replaced.

Ms. F's last face-to-face lecture to her students was a lesson in what works. Safety and science must be inextricably integrated.

3

Connections

- American Society for Microbiology hand washing information
  www.microbe.org/washup/wash_up.asp
- ANSI Z87.1, American National Standards Institute
  www.ansi.org
  www.coltslaboratories.com/
- Environmental Protection Agency IAQ INFO 800-438-4318
  www.epa.gov
- FDNY New York City, New York Fire Department
  www.nyc.gov/html/fdny/html/safety/extinguisher/classes.shtml
- Fume Hoods Fact Sheet, Office of Environmental Health and Safety, University of California, Berkeley, 11 September 2003
  www.ehs.berkeley.edu/pubs/factsheets/09fumehd.html
- Laboratory Safety Institute
  Designing, remodeling, and building safer labs. Natick, MA:
  www.labsafety.org
- MSDS (materials safety data sheet)
  www.flinnsci.com
  www.esf.uvm.edu/uvmsafety/labsafety/chemsafety/netmsds.html
  http://msds.ehs.cornell.edu/msdssrch.asp
- National Fire Protection Association
  www.nfpa.org/index.asp
- National Institute for Occupational Safety, Chemical Safety
  www.cdc.gov/niosh/topics/chemical-safety
- NFPA 45, Standard on Fire Protection for Laboratories Using Chemicals
  www.nfpa.org/Codes/NFPA_ Codes_and_Standards/List_of_NFPA_documents/NFPA_45.asp
- NFPA 101, Life Safety Code
  www.nfpa.org/BuildingCode/AboutC3/NFPA101/nfpa101.asp
- NFPA 5000, Building Construction and Safety Code
  www.nfpa.org/catalog/product.asp?pid=500003&src=nfpa&link_type=buy_box&order_src=A292
- Occupational Safety and Health Administration
  www.osha.gov
- OSHA, Hazard Communication
  www.osha.gov/SLTC/hazardcommunications/index.html
- OSHA Regulations (Standards—29 CFR) "Laboratory Standards"
  Occupational exposure to hazardous chemicals in laboratories. 1910.1450.
  www.osha.gov/pls/oshaweb/owadisp.show_document?p_table=STANDARDS&p_id=10106
- OSHA Regulations (Standards—29 CFR) "Chemical Hygiene Officer and Program"
  National Research Council Recommendations Concerning Chemical Hygiene in Laboratories (Non-Mandatory). 1910.1450 App A.
  *www.osha.gov/pls/oshaweb/owadisp.show_document?p_*table=STANDARDS&p_id=10107
- SIRI, Vermont Safety Information Resources, Inc.
  www.siri.org

Finders Keepers

Essentials of Safer Storage

4

More than any other community college instructional disciplines, laboratory sciences require safe and secure storage facilities. This includes storage within the classroom/laboratory and also dedicated storage areas adjacent to but separate from the classroom/laboratory space. Safety in science materials storage rests on three issues:

- *Limiting the amount stored.*
- *Limiting the type of materials stored.*
- *Limiting access to the storage spaces.*

The Stuff of Science

Whether we find ourselves in a building old enough to be historically significant or in newly constructed quarters, there never seems to be enough storage space for the typical science department. It may be our insecurity about long-term budgets. It may be the constant time crunch, which can keep us from organizing and thinking through what we put away. But we always store more than we need. It doesn't take long for the accumulated materials to overflow storage areas, take over valuable counter surfaces and start blocking access to safety equipment, aisles, and exits. Even worse, the chemicals and materials that science instructors hoard can deteriorate and become dangerous.

Properly managing and storing all the "stuff" needed for an effective laboratory program requires both knowledge and discipline. It requires thorough knowledge of the materials and their potential for harm. It also requires instructors to plan like generals and cooperate like roommates, while maintaining a level of security the local bank would envy. Then there is the most difficult discipline of all—keeping only what is absolutely necessary.

A Place for Everything

Well-designed science facilities have cabinets that are varied and specialized—fireproof cabinets; chemical-resistant surfaces; long, flat drawers for posters; and tall, narrow places for graduated cylinders and microscopes and everything in between. But science classrooms should not be the only places where the materials for laboratory work are stored. Separate, secure science storage rooms and prep rooms are also required. Storage and prep rooms should be situated as close to point-of-use as possible, preferably with direct access to the laboratory so that materials won't be transported through the corridors. Each storage area should have separate keycard or keypad access that will reveal who is entering and that will block access for people who no longer need to be there.

Chemical Storerooms

Storerooms need special features: specialized storage cabinets, special air handling, and emergency equipment. Chemical storerooms should not be combined with preparation rooms or offices. That would pose a serious hazard to the instructor. Even with extra ventilation, the air in chemical storage rooms is likely to have toxins in quantities that may not be detected but could prove harmful with prolonged exposure. If an accident occurs during preparation and you are working in a chemical storage room, the proximity of stock quantities of chemicals could be disastrous.

If a chemical stockroom has only one entrance, it must be easily accessible. Keep the door area clear of the most hazardous and least stable chemicals. The chemical storeroom should have highest priority for upgrades and improvements.

Prep Rooms

The prep room has a different purpose than a storeroom. Well-equipped spacious prep rooms are especially valuable if instructors do not have access to their own classrooms for preparation prior to meeting with students. Include space for lab setups to be locked up safely if the classroom or laboratory is inaccessible or unsecured. Prep rooms should have emergency equipment such as showers and eyewashes as well as separately lockable storage cabinets—mobile or not—to secure materials that have been prepared.

SCILINKS THE WORLD'S A CLICK AWAY
Topic: chemical safety
Go to: *www.scilinks.org*
Code: SSCC50

Prep rooms should not be used as offices or meeting rooms. Although we all need the company and camaraderie of our fellow department members, the prep room is not the place to gather, have lunch, or hang out. When missions are mixed, such as in using a prep room for department meetings or lunch gatherings, the original intent is blurred. Is this a place of science with inherent dangers to be kept in mind or a lunch counter and social room? Keep that distinction for the safety of all. A prep room is a science facility and only that.

Anchoring

Many state and local building codes require that nonmovable cabinets be anchored to the wall so they do not cause injury in a tipping accident. Even if there is no such code in your jurisdiction, check all your cabinet and storage units for proper stability and insist on having the units anchored to the wall.

Movable carts may be prone to tipping, especially tall ones such as those used for large televisions or video displays. Be sure to lock and anchor movable equipment carefully and check for stability before attempting to move them.

Federal, state, and local building codes have seismic requirements—even for areas the public may not think of as earthquake-prone. You should consider meeting the requirements even if your facility is grandfathered, because bumping and shaking could occur accidentally even without an earthquake.

And Never Your Lunch!

Refrigerators in the laboratory, science storeroom, or science prep room must never be used to store food or drink for human consumption and should have prominent signage: No Edible Food Storage. If the refrigerator is ever used to store toxic chemicals, then food for classroom animals should not be stored in it either.

Any refrigerator used to store combustibles must be a spark-free, explosion-proof laboratory refrigerator—not an ordinary kitchen refrigerator that can generate a spark every time it begins to run. Science rooms are not kitchens, figuratively or literally—no exceptions.

4

Safe and Secure

Both the value and hazards associated with science equipment and supplies dictate that special consideration be given to securing work spaces and materials. Instructors must consider not only how materials and supplies are used for instructional purposes, but whether, in the wrong hands, they can be used for malicious or criminal activities. Items such as potential explosives, raw materials for drugs, radioactive sources, and infectious organisms may have dual or multiple uses. You need to consider whether the value of having and storing these materials for instructional purposes is worth the risk of legal and ethical liability should they be stolen. For those items you must have, security is a critical element of a safe storage system.

Locks and Keys

All science-related rooms—labs, storage rooms, and preparation rooms—should have a lock system that tracks access and allows entry only by authorized instructors and specific custodial staff. A keycard system works best for this purpose. Each user is issued a unique card. The time and date of each entry can be matched to individual cardholders. The system administrator can render cards inoperable when an individual leaves or no longer needs access. A keypad with a revisable code may work better than a regular key system, but users must be careful that they are not observed while punching in the code.

Storage rooms and preparation rooms should each have separately locked doors with keying different from lab, classroom, and nonscience storerooms. No one should be able to enter a science storage or prep room just by using a regular classroom key.

4

LOCATION, LOCATION

Sometimes sensitive electronic equipment and delicate instruments are stored on a cart and moved from a storage place to point of use. If the regular storage space for the loaded cart is in a storeroom or prep room, be sure that the space is not near chemical storage. Fumes from a chemical storage room or area can quickly corrode delicate mechanical movements and electrical circuitry.

Electric and electronic equipment (except specifically designed and protected probes) should not be anywhere near water. Goosenecked sink fixtures too easily become a source of spray. Moisture and electronics just do not mix, and most commonly used laboratory equipment is not moisture proof.

Exercise care when using equipment in a marine environment. Whether at a field site or in a classroom simulation, saltwater is corrosive. Microscopes are particularly at risk from corrosive moisture that finds its way below the stage and into moving parts.

Lights On for Safety

For storage and preparation rooms that are not visible when doors are closed, we recommend installation of a signal light at the entrance door that remains lit whenever the lights inside the storage or prep room are turned on or a motion detector is activated. In case of emergency, this will alert passersby or emergency personnel that someone is working inside the room. (An added value of an electronic key system is that the facilities

(cont. on next page)

Access by Whom?

Access should be restricted to individuals with specific need for the access. Only the well-qualified laboratory coordinator or department head needs access to all the storerooms. Individual instructors should have access only to the areas and materials they need and only during the period of their assignment.

In some configurations, a single preparation area is directly connected to two or more labs. If this is the case, specific steps should be taken to prevent students and others from cutting through the prep area to get from one place to another. If the prep area is used as a passageway, there is no accounting for who was there. Loss of supplies and equipment and contamination of stock solutions and chemicals are just a few of the problems that can result.

At the college level, credited students or paid lab assistants are the norm. It's important to remember that the instructor retains major responsibility for what is taken from the storeroom, what is prepared, and what is put away or disposed of after the laboratory activity is done. The entire department, including part-time instructors, should be involved in establishing work rules for laboratory assistants. No matter what their training, lab assistants should work only on preparations when there is a responsible faculty member available for help and support.

Students should never be permitted to enter storage rooms. Neither should they be permitted to enter or work unsupervised in prep rooms—not even your most reliable, responsible students.

(cont. from preceding page)

manager or security officer can get a signal if someone is in a storeroom or prep room at an unexpected time.) Also recommended are emergency or "panic" buttons inside prep and storage areas to be used in case of accident or emergency.

When instructors and support personnel leave or are reassigned, access to storage and preparation areas should be checked, keys and card passes returned, and other means of access eliminated. Substitutes may be given temporary access only if properly trained.

A REVOLVING-DOOR FACULTY

With full-timers and part-timers, permanent and adjunct staff, turnover of community college science faculty from semester to semester can exceed 50%. Additionally, the background, experience, and training of faculty members can vary widely. It's not good policy to give the toughest teaching assignments to the most junior faculty, especially if the instructor has to manage, access, and store a vast array of materials and equipment with which he or she may have only minimal familiarity.

Materials and equipment essential for one instructor's course may be completely unfamiliar to another instructor and can pose inadvertent hazards to colleagues and other staff. For this reason, we recommend that storage areas be divided as much as possible so that individual faculty members can store and lock supplies for their specific courses within any larger classroom or storage area.

Upon departure, faculty members should be asked to properly dispose of and clear out all expendable materials for their courses and return all keys. Equipment and nonexpendable items should be reviewed by the department chair and either re-assigned or appropriately disposed of if not likely to be used within the next two years.

4

For Your Own Safety

There have been many accidents where instructors working alone have been unable to get quick help. For your own safety, we recommend that you never work alone in a laboratory, storage, or prep area. In case of accident, there should be someone available to help or to summon help. We highly recommend a phone or panic button.

Storage in the Classroom

Classroom storage may be the most convenient. It is tempting to put everything you might need in easy reach, but you should consider what is appropriate to have in the classroom and how storage units are distributed around the classroom. Some materials may be attractive nuisances—better to keep them out of sight and out of mind.

Strategic Locations

In-class storage may be provided for frequently used labware such as glassware, clamps, tubing, cylinders, and filter paper. This type of storage should be distributed throughout the room, occupying space on at least three of the four walls so that students can access the supplies without interfering with one another. Use lots of labels and organizational diagrams around the room and on classroom cabinets so everyone can see where everything is kept and help with the distribution and storage of most equipment.

You might also use tote drawers or dish tubs to store and distribute all items needed for a particular activity or for each student work team. Include a laminated card with a list of the supplies and safety precautions. This type of system can be very convenient for lab setup and cleanup and for visual learners and absentees. Post a digital photo or diagram on the inside of cabinet doors showing where everything belongs.

Out of Sight

Some useful items are too dangerous or tempting to be stored in the classroom. Although it may be convenient to have equipment stored in the classroom, you should also consider whether the security in the room can protect against loss of expensive instruments. Balances, in particular, should not be left out. They have very high street value and are among the most frequently stolen items from laboratories. If equipment is so frequently used or is so sensitive that it is unreasonable to store it elsewhere, then consider having it locked or bolted to permanent fixtures in the room. Give similar consideration to computer equipment, particularly laptops.

Chemical stocks should never be stored in the classroom. Transfer the amount of material needed for the activity into a properly labeled container or containers for use in the classroom. This protects the stock from contamination. (Will your students remember to only pour out and never place anything into the stock bottle?) The practice also reduces the potential harm should the chemical be reactive or be stolen.

If you wish to further subdivide chemicals for distribution, remember that smaller containers must have the same labeling information as the stock bottle. The labeling effort is a good assignment for paid or credited student assistants. Computer-generated labels can include hazard icons as well as material safety data sheet (MSDS) information that students should be taught to recognize and understand.

Through the Looking Glass?

Many science classrooms are equipped with glass-door cabinets. Everything needed for classes and often some things not needed are there for all to see. But experience has shown that glass doors are not a good idea. They contribute nothing to security and are too easily broken. Things stored behind the glass tend to get messy and can become significant distractions to students.

No More Minimuseums

4

Jars of preserved materials were once considered objects for display. Today, many of these souvenirs are considered toxic, politically incorrect, or just disgusting. Why do we keep the one-eyed "Cyclops" pig in its two-quart jar of formalin? Is it a valued teaching and learning tool or a "yuk" generator, something to shock the students? If it has a strong academic purpose or a critical role in the program, keep it. If it does not, get rid of it.

DESIGN YOUR OWN

If you can help in designing or remodeling a science classroom, be sure to build safe storage into the walls. Include many lockable cabinets of various types and drawers with slots for labels. If several different departments or faculty members with different course assignments share the same classroom, separate keys for different storage areas are important.

Shallow cabinets can be built behind the chalkboard or projection area, behind bulletin or chart boards, and in other spaces to maximize your ability to keep the materials you need close at hand. Shallow cabinets work better than deep cabinets, because stored items can be reached more easily and you avoid getting things shoved back where they can never be found again.

Include flat storage drawers to store posters and charts, vertical storage space for tall items such as metersticks, and larger, low spaces for models and bulky equipment.

Choose cabinets and equipment that have rounded rather than sharp corners and edges. Strong, well-reinforced wood cabinets work well while plastic

(cont. on next page)

(cont. from preceding page)

laminates may be less durable. Metal cabinets tend to have sharp corners and edges and may become corroded.

Shelf storage should have lips or dowel fences along the outer edge to prevent items from falling off. Check seismic requirements that are part of federal, state, or local building codes.

4

Chemical Storage

Chemical storage is an issue for all science instructors, not just the chemistry instructor. A good storeroom has a minimum of chemicals. Chemicals should be purchased in quantities that will be used up in the year purchased (or, at most, two years).

The most important thing to remember is that *less is better*—less quantity, lower concentrations, and less handling.

Material Safety Data Sheet (MSDS)

A material safety data sheet (MSDS) is a standard document available for every hazardous chemical manufactured or sold in the United States. The MSDS contains information in a specific format so users and emergency personnel can find needed data expeditiously.

Federal and many state laws require that MSDSs be available *all the time* for *every chemical* you use or store. You may be surprised to learn that *every chemical* includes everyday items such as detergent, hand soaps, and household bleach. Administrative and noninstructional staff, instructors, students, and emergency responders must have easy access to MSDSs for materials stored and in use.

For practical purposes, there should be at least two sets of documents. A complete set of MSDSs for everything in the institution should be kept in a central location while duplicate sets should be located wherever the substances are used. Your local fire department should be provided with an inventory and map of materials stored and the location of the MSDS files for the materials. First responders need to know what they might encounter in a fire or other emergency.

When chemicals are ordered through a purchase order system, the requisition should specify that the vendor must "Provide MSDS with order. Shipment must have MSDS to be accepted." The preferred method for meeting the MSDS requirement is to train whoever is responsible for receiving and accepting packages to look for the MSDSs before accepting the shipment. If the required MSDSs are not in the shipment, the shipment should be refused—not held in an office or storeroom awaiting proper documentation. Once the package is accepted, copies should be made of the MSDS documents—one for the master file and enough duplicates to accompany supplies to each separate storage and use area. Remember that accepting a shipment without an MSDS and holding up payment does not fill the requirement of having an MSDS on file for every chemical present in the institution.

The need for MSDSs is just one reason that part-time instructors should *never* transport materials from one job or assignment to another. College materials should not be borrowed for high school projects or vice versa.

When a common supply such as detergent is available for instructors to take from central storage, there should be multiple copies of its MSDS available to take along with the supply. If you take supplies from a central location, you need to take a copy of the MSDSs along for your own MSDS file. The location of MSDS files should be readily apparent in every room.

If you need to bring chemicals from home or purchase them at a convenience store because they are less expensive, you must obtain the MSDS yourself. You can sometimes find a telephone number on the bottle. A good source of MSDS information for common household products is the Vermont Safety Information Resources, Inc. (SIRI) site. (See "Connections" at the end of this chapter.) Here's an example of a common material and the standard MSDS that you can obtain for it.

MATERIAL SAFETY DATA SHEET—BLEACH LAUNDRY ORGANIC CHLORINE

(Abbreviated version)

General Information

Item Name: BLEACH LAUNDRY ORGANIC CHLORINE

Company's Name and Address: ______________________________

Company's Emerg Ph #: ______________________________

Ingredients/Identity Information

Proprietary: NO Signs/Symptoms Of Overexp: INHALATION: IRRITATION TO NOSE, THROAT, MOUTH SEVERE IRRITATION AND/OR BURNS. EYES: SEVERE IRRITATION AND/OR BURNS CAN OCCUR.

Emergency/First Aid Proc: EYES: FLUSH W/LG AMTS WATER—15 MIN, OCCASIONALLY LIFT UPPER/LOWER LIDS. CALL DOCTOR. SKIN: FLUSH—15 MIN. CALL DOCTOR. REMOVE CONTAM CLOTHING AND WASH BEFORE REUSE. INGEST: DON'T INDUCE VOMIT. DRINK LG AMTS WATER. DON'T GIVE ANYTHING BY MOUTH IF PERSON IS UNCONSCIOUS OR HAVING CONVULSIONS. INHALE: REMOVE TO FRESH AIR. IF BREATH HARD, GIVE OXY. KEEP WARM/REST. IF BREATH STOPS, GIVE CPR. CALL DOCTOR.

(cont.)

MATERIAL SAFETY DATA SHEET—BLEACH LAUNDRY ORGANIC CHLORINE

(Abbreviated version, cont.)

Precautions for Safe Handling and Use

Steps If Matl Released/Spill: IF SPILL IS DRY, CLEAN UP W/CLEAN, DRY DEDICATED EQUIP & PLACE IN CLEAN, DRY CNTNR. SPILL RESIDUES: DISPOSE OF AS NOTED BELOW. NEUTRALIZE MAT'L FOR DISPOSAL. CALL 1-800-654-6911
Waste Disposal Method: PRODUCT DOES NOT MEET CRITERIA OF HAZARDOUS WASTE. AS A NONHAZARDOUS SOLID WASTE, DISPOSE OF PER LOCAL, STATE, & FEDERAL REGULATIONS BY TREATMENT IN A WASTEWATER TREATMENT CENTER.

TAKE CARE TO PREVENT CONTAMINATION FROM THE USE OF THIS PRODUCT. COOL, DRY, WELL-VENT AREA. DO NOT STORE ABOVE 140 F OR IN PAPER/ CARDBOARD. KEEP CLOSED & FROM MOISTURE.
Other Precautions: ADDITIONAL RESPIRATORY PROTECTION NECESSARY WHEN SMALL, DAMP SPILLS INVOLVING PRODUCT OCCUR, WHICH RELEASES CHLORINE GAS. FULL FACE CARTRIDGE-TYPE NIOSH APPROVED RESPIRATORY W/CHLORINE CARTRIDGE RECOMMENDED. USE SELF-CNTND BREATHING APPAR.Ingredient: SODIUM CHLORIDE
Ingredient Sequence Number: 01
Percent: 45-50

Proprietary: NO
Ingredient: SODIUM TRIPOLYPHOSPHATE
Ingredient Sequence Number: 02
Percent: 25-30 Proprietary: NO
Ingredient: SODIUM DICHLORO-S-TRIAZINETRIONE
Ingredient Sequence Number: 03
Percent: 24-28

Fire and Explosion Hazard Data

Extinguishing Media: USE MEDIA TO CONTROL A SURROUNDING FIRE. DO NOT USE DRY CHEMICAL EXTINGUISHERS CONTAINING AMMONIUM COMPOUNDS.
Special Fire Fighting Proc: USE WATER TO COOL CONTAINERS EXPOSED TO FIRE. SMALL FIRES—USE WATER SPRAY OR FOG. LARGE FIRES—USE HEAVY DELUGE OR FOG STREAMS.
Unusual Fire And Expl Hazrds: NONE. REQUIRED BEFORE EXTINGUISHMENT CAN BE ACCOMPLISHED. THE USE OF SELF-CONTAINED BREATHING APPARATUS IS REQUIRED IN A FIRE WHERE THIS PRODUCT IS INVOLVED.

(cont.)

MATERIAL SAFETY DATA SHEET—BLEACH LAUNDRY ORGANIC CHLORINE

(Abbreviated version, cont.)

Reactivity Data

Stability: YES
Cond To Avoid (Stability): ELEVATED TEMPERATURES (ABOVE 140 F)
Materials To Avoid: OTHER OXIDIZERS, NITROGEN CONTAINING COMPOUNDS, DRY FIRE EXTINGUISHERS CONTAINING MONO-AMMONIUM PHOSPHATE.
Hazardous Decomp Products: NITROGEN TRICHLORIDE, CHLORINE, CARBON MONOXIDE.

Health Hazard Data

Route Of Entry—Inhalation: YES
Route Of Entry—Skin: YES
Route Of Entry—Ingestion: YES
Health Haz Acute And Chronic: INHALATION OF HIGH CONCENTRATIONS CAN RESULT IN PERMANENT LUNG DAMAGE. CHRONIC INHALATION CAN ALSO CAUSE PERMANENT LUNG DAMAGE. SKIN: PROLONGED EXPOSURE MAY CAUSE DESTRUCTION OF THE DERMIS WITH IMPAIRMENT OF SIGHT TO PROPERLY REGENERATE. EYES: MAY CAUSE VISION IMPAIRMENT AND CORNEAL DAMAGE.

4

Organizing Stored Chemicals

Chemicals should be stored by reactivity and compatibility—rather than alphabetically—in cabinets and on shelves specially designed for the purpose.

- Cabinets for the storage of flammables should be made of materials that do not burn. Most are made of special polymers that can delay, but not prevent, a fire. There is disagreement regarding whether a storage cabinet for flammables should be vented. Some believe it is appropriate to vent the cabinet to prevent the accumulation of fumes, but others believe venting introduces extra oxygen that may actually feed a fire. There is no federal requirement to vent these cabinets. Tightening caps and sealing them with vinyl electrician's tape is a good practice.
- Acids and bases should be stored separately from each other in corrosion-resistant cabinets that ideally are ventilated to the outside, above the roofline, and away from air intakes. These acid-storage and base-storage cabinets are commonly made of wood, although some plastics work well. These cabinets should never be stacked on other cabinets. Be cautious when you purchase them. Some cabinets supposedly made for storage of corrosives have been found to have metal supports, which can corrode and collapse.

- Potentially reactive chemicals should not be clustered together. For example, separate acids from bases, and oxidizing acids such as nitric and perchloric from organic acids such as formic and acetic. There are many potentially risky combinations of chemicals such as oxidizing agents close to reducing agents. Be sure you check MSDSs and other references carefully.
- Be alert to crystalline or filmy coatings on container or cabinet surfaces. They are usually signs of leaking containers or reactions among reagents.
- Be especially careful to record and monitor the age of reactive and unstable chemicals.
- Some reactive chemicals should be stored in a container within another container or bucket in case the inner container is broken.
- Toxic chemicals require not only secure storage but also specific, and expensive, disposal procedures. Consider less toxic alternatives and minimizing quantities acquired.
- Open shelving for the more benign chemicals should be sturdy and solid, with short spans between supports. There should be lips or dowel fences along the edges to hold items in place in case of shaking or bumping. Shelves should be shallow, no more than 20 cm deep or two bottles deep.
- Avoid shelves placed too high. Labels are hard to read, and the bottles are more likely to tip or be dropped when being removed.
- Store heavier bottles on lower shelves. Be careful to consider potential reactivity if something stored above should fall on something below.
- Keep the doorway area clear and unobstructed. Bottles and jars hastily or "temporarily" placed near doorways, carboys on the floor, and a pile of newly arrived supplies are a bad combination. Keep your exit route clear.

These are basic principles for the design of your chemical storeroom. But just because you now know how to store corrosive, reactive, flammable, and toxic chemicals does not mean they should be there. The laboratory programs of 10 and 20 years ago required more and stronger chemicals than what are acceptable today. If your chemical stores were amassed then—or if you inherited materials from another institution—you may have many more chemicals in your storeroom than you need. That means it's time to call a reputable hazardous waste disposal firm and clear out unnecessary materials.

Low-Level Exposure

Stored chemicals releasing low concentrations of vapor may never be noticed by students or other temporary occupants but could trigger an allergic reaction or worse in an instructor or lab assistant who occupies the room for extended periods. Chronic allergies and other conditions may be related to extended low-level exposure to environmental chemicals.

If your program requires acids, bases, or other potentially corrosive chemicals, it is better to purchase smaller, prediluted quantities and consider alternatives to lab activities that require strong chemicals. Review all of your lab activities to see if they can be done in microscale. A microchemistry approach uses small amounts of chemicals, droppers, and clear plastic plates with test wells that reveal changes clearly and wash up quickly.

AGE DISCRIMINATION

Chemicals are not like fine wines. They do not grow better as they age. Many are reactive, unstable, or decompose to become very dangerous. Consider just a few examples:

- Ethers, highly volatile and flammable to begin with, can react with oxygen in air to form peroxides so unstable that just turning the cap can cause an explosion.
- Picric acid was once used to preserve specimens and for biological staining. However, it becomes unstable and explosive when dried out and should not be used. If you should find picric acid or suspect something could be picric acid, do not touch or move it, and evacuate the area immediately. Just jarring it can cause it to explode. Call immediately for professionally trained help with removal and disposal.
- Many powdered reagents become so hard they cannot be removed from their stock bottles. Attempting to break up a chunk can break the bottles, cut your hands, scatter the chemicals, and possibly result in explosion.

Date every purchase. Buy only what you need for one year, even if it's cheaper to buy in quantity. Consider the costs and hazards of disposal before you purchase. Properly dispose of all excess and outdated chemicals at the end of each academic year. Your course might change, program needs might be different, or you might have another assignment. Tell the administration about this important principle. They should budget for new consumables and licensed hazardous waste disposal each year.

Disposal and Cleanup

Disposal is complex and can be very expensive. Look up the disposal issues *before* you purchase something. Some recommendations of reputable national organizations and agencies have been questioned by local officials, so make sure you check local regulation as well as state and federal standards. If it's too hard to

SCiLINKS
THE WORLD'S A CLICK AWAY
Topic: hazardous waste
Go to: *www.scilinks.org*
Code: SSCC61

4

4

No Dumping

You cannot dispose of most unwanted chemical and biological waste by pouring the materials down the sink, flushing them down the toilet, or dumping them into the trash. Most of your unwanted materials can present hazards to the people who must handle them next and environmental hazards when they enter the waste stream or watershed. Your excess chemicals do not belong in any of our waterways. Some chemicals may be diluted or denatured and then discarded, but it is vital that the directions for these procedures be followed to the letter.

get rid of, don't buy it. Request that the institution or department establish a contract with a reputable hazardous waste disposal company to remove hazardous wastes from all locations.

The cost of disposal is one more reason why chemicals should *never* be borrowed or lent between institutions. If a researcher from industry brings a chemical to his part-time adjunct instructional site, or a college instructor lends materials to a high school teacher, the liability and disposal costs come along for the ride.

Some materials that are innocuous in small quantities can do extensive environmental damage when they go down the drain. One example is phosphates, found in strong detergents and plant foods. Significant quantities in small bodies of water cause eutrophication and damaging algal growth. Animal droppings from cage bedding can have the same effect. Even common salt can be a problem. Be cautious of dumping anything down the drain. Disposal rules are constantly changing. Be sure you know before you decide to dump anything.

You are responsible for the safety of everyone who handles the waste from your classroom, including custodial staff and trash collectors as well as anyone who may use your classroom or preparation area after your activities.

BEWARE THE GEEK BEARING GIFTS

To save yourself from liability, responsibility, and a lot of expensive and unrewarding work, never accept gifts or donations of chemicals from research labs, business and industry, or anyone else. You will not have the appropriate MSDSs, and you cannot be certain of the age, purity, and prior storage conditions of chemicals that are not ordered and received directly from a reliable science supply house. You may even find that some of your "gifts" are actually hazardous wastes for which the donor did not want the responsibility of disposal. Some materials not subject to regulation when you first accept them may later be declared hazardous. The responsibility and cost for hazardous waste disposal becomes yours.

YOURS FOR LIFE

Did you know that from the moment you purchase a chemical, your institution is responsible for its disposal?

Even if you turn it over to someone else, if the subsequent owner stashes it in the wrong place or disposes of it improperly, the chain of responsibility can bring the liability back to you and your institution.

Even if you contract with a legitimate hazardous waste disposal company, you could still be encumbered with additional costs. If the company makes a mistake that results in contamination and as a result goes out of business, liability can rebound to the company's customers.

Chemical Waste

For chemicals you have now, look up the proper disposal methods or seek professional assistance. Heavy metals such as cobalt, nickel, mercury, and lead should never be put into the wastewater system or trash; neither should herbicides and pesticides. Simultaneous disposal of acids and bases and other combinations can cause an explosion in the pipes or traps that can cost thousands of dollars to repair.

Old boxes, bottles, carboys, and other containers containing chemicals that no longer have legible labels should be treated with great caution. If you cannot be sure of contents, then you must assume the materials to be hazardous. Professional assistance is needed.

Are You a Toxic Waste Generator?

A complex system of environmental laws and regulations governs toxic waste disposal. Depending on the total mass of toxic waste generated annually, your institution may be classified as a "large waste generator," which requires special permits and significant expense. The waste includes not only your materials but also wastes from other departments such as glazes and acids from the visual arts program and cleaning solutions the maintenance staff uses. If the toxic chemicals are in solution, the total mass of the solution is counted.

At Your Disposal

- Be sure you read the MSDS when planning for proper disposal.
- Find out about federal, state, and local regulations.
- Your local fire department may be willing to take some hazardous wastes to use in training and emergency drills.
- If in doubt, consult with those who know about disposal, such as the department chair, facilities manager, insurance risk manager, hazardous waste coordinator, fire chief, and the state or federal Environmental Protection Agency.
- Every institution or community should have a hazardous waste coordinator (aka a chemical hygiene officer) who is knowledgeable about federal, state, and local regulations.

Old Favorites

We used to routinely have our students type their own blood, use saliva, and test urine. Those days, if not past, are almost gone. We must re-examine our practices because we now recognize deadly diseases are passed by body fluids. Laboratory exercises using human body fluids are best ended and replaced.

4

Do not stockpile your waste from year to year. Request that your institution contract with a responsible hazardous waste disposal firm to remove hazardous wastes on a timely and regular basis. At the same time, limit your acquisition and use of such material. Investigate alternatives and experiments that require smaller amounts of chemicals.

Biological Waste

Organic matter must be sterilized or otherwise decontaminated before it is discarded. If you grow cultures, they must be sterilized before disposal. Flood agar plates with 10% chlorine bleach for a minimum of five minutes. Kill molds with bleach or another disinfectant. Wrap terraria and animal bedding in plastic. If you use autoclaves, make sure that they are safely installed in a secure location and that instruction for proper use is strictly followed.

Do not keep preserved specimens for more than a year. Molds grow even in preservatives. If you have formaldehyde or specimens preserved in formaldehyde from years ago, you cannot simply throw these materials away or dump the fluids down the sink. They are considered hazardous wastes and must be handled as such.

Know the institution policies and procedures for handling human body fluids. (See "Standard (Universal) Precautions," Chapter 10, pp. 156–157.) Make sure anyone who might come into contact with body fluids is prepared to use Standard (Universal) Precautions. Students should be given an abbreviated set of rules consistent with institution and public health policy.

NATIVE AMERICAN HUMAN REMAINS

When you were hired 23 years ago, there was a box in the back room that was covered with dust and tightly sealed. Attached to the box was a note indicating its contents had been removed from the glass display case at the back of the room. At long last, curiosity prevails, and you open the box. Bones—human bones!

The explanation could be quite simple. There was a time when bones unearthed through construction or simple soil renovation projects were brought to institutions for use as displays. They will most likely be of no archeological

(cont. on next page)

(cont. from preceding page)

significance since their origin will have been lost. But some of these bones, including skulls, may be Native American, dating back hundreds if not thousands of years.

Native Americans are very concerned that their ancestral remains, even fragments, be handled with the greatest respect. If you believe you may have Native American ancestral remains, first contact your department head. Your local coroner or university should be able to determine if they are Native American remains. If they are determined to be Native American ancestral remains, your institution should send a formal letter to the tribal chairman of the closest Native American tribe. Tribal officials will come to your school and remove the bones. They may conduct a quiet ceremony. It is important that the bones be treated with the utmost respect at all times. It would not be a positive action to dump the bones, dirt, spiders and all on a back room table awaiting tribal members. Keep them together and secure in a closed and clean cardboard box or other suitable container.

Should the bones be determined not to be Native American, the local coroner's office should handle the matter. A quick and responsible reaction can resolve this issue with respect and common sense.

4

Equipment Waste

Disposing of equipment can also be difficult. You need a special container for sharps such as blades, broken glass, and metal pieces. You must see that this container is removed in a proper way on a regular basis. See Chapter 10, pp. 157–158, for more information on sharps.

Instruments containing mercury or other heavy metals must be handled as hazardous waste. There are also special disposal procedures for electronic equipment, cathode ray tubes, fluorescent tubes, smoke detectors, dry cells, and other items.

Itinerant Instructors and Shared Spaces

Itinerant instructors moving from room to room (science on a cart) should be discouraged. This practice is not safe. Also to be discouraged is shared lab space when no single person can be held responsible for the conditions in the room. Schedules that do not provide instructors with access to their rooms prior to the scheduled class time are also bad practice. These conditions greatly increase the chance for accidents.

4

Your Partners, the Maintenance Staff

Many instructors believe that once something is thrown away, it will not trouble them. Think again! You are also responsible for the health and safety of your maintenance staff and others who might come in contact with lab waste. Make the custodians informed members of your teaching team. They too need to know about your laboratory activities and how they may affect their duties. Develop safe disposal rules and enforce them with colleagues, lab assistants, and students.

Itinerant instructors can never be sure of what may have been left out from the class before or what may linger on a desk or counter. When one instructor leaves a room and the next instructor has not arrived, the laboratory may be left unsupervised, creating a hazardous situation especially if any materials are left in the open between sessions.

Instructors and department chairs should work together, take a hard look at the program, and make the modifications necessary to ensure safety. If any instructor must use multiple rooms, it should be the most experienced full-time instructor that does so. Whatever the conditions and the schedules, safe and secure storage must be provided for lab supplies and equipment—even if some of this storage is in the form of lockable portable units. The administration must understand that there is potential for institutional liability if chemicals and specialized equipment are not properly stored and secured.

If you must change rooms during the day or teach a class in the evening when full-time staff is not present, ask for lockable storage that is exclusively for your use. If you are an adjunct instructor or sharing your lab space with another instructor, make sure there is an ongoing system for sharing information about storage and safety issues and a day-to-day communication system to keep each other informed about special projects. It is important that all instructors sharing the space come to clear agreement on all rules and all procedures.

Briefcases, Backpacks, Outerwear, and More

Students' briefcases, backpacks, and totes can create safety and security problems. Ask students to bring minimum baggage to laboratories and to be prepared to stow all but the supplies needed to complete the lab exercise. Arrange a storage area away from counters, tables, and other laboratory work areas, preferably outside the lab, where everything extra can be kept during labs.

If student laptops or calculators will be used during lab work, make sure these items are clearly marked with the owner's name. Electronic equipment such as cell phones, MP3 players, and personal data assistants (PDAs) should be in off mode and stowed during class.

Arrange a place away from work areas where outerwear can be hung during class. Do not permit jackets, purses, backpacks, or other such items to be hung on the backs of lab stools or chairs where their weight can cause furniture to tip.

Give This Book Away

Key issues raised in this book, particularly those related to facilities and storage, may not be issues that an individual instructor can address. Even the department chair may not have sufficient authority to address the need to provide safe and secure premises for laboratory work.

You may have inherited major storage and disposal problems and need a department meeting or meeting with facilities managers to arrange for solutions. At the meeting, concentrate on teamwork. Don't waste time and energy blaming former staff. Consider providing nonscience administrators and facilities managers with annotated copies of this book or copies of relevant sections of this book to make them aware of their responsibilities and liabilities to provide safety equipment and safe facilities.

Remember that a seemingly simple request to clean up old and excess supplies may require longer-range institutional planning. When budgets are tight, funds may be unavailable right away but should be reserved in the next budget round. Plan for the future. You may need a special allocation of funds or the services of a professional environmental hazard disposal firm to make your storage facility safe again. Once you have legally disposed of old chemicals and equipment, make sure they don't accumulate again.

Absence Makes the Lab Grow Sounder

Whether you are moving into renovated or new quarters, or just returning to old ones, you should take a hard look at your inventory each year. Consider the program needs for the coming year, and keep very little else. Excellent college programs can function with limited storage, but it takes knowledge, care, and constant vigilance to do so.

Rule 1: Anything you haven't used in two years you probably don't need. And, realistically, if you decide you do need it, you probably won't be able to find it and will buy it again.

Rule 2: Never accept hand-me-down chemicals and most specialized equipment from another institution, your other job, or industry. Chemicals have shelf lives, and hand-me-downs may be of uncertain age and purity. The contents may have been contaminated by use. As the new owner, you will become responsible for disposal, which can be very costly.

Rule 3: Date your purchases. (See "Age Discrimination," p. 61.) If you have access to chemical database software, enter each purchase. At the very least, mark the date of purchase with indelible marker on the container and in the MSDS notebook.

Rule 4: Get a good storage guide for everything you buy. Catalogs and information from science supply companies often have safe storage and handling information.

Rule 5: If you can reduce the amount of stored items, do so. Scan worksheets and store them electronically. Keep a URL list rather than catalogs for supplies. Almost all companies have online catalogs now.

Rule 6: Never ignore a spill or a cloudy film on the inside of a chemical-storage cabinet. They are clues that the chemicals you have stored may be reacting or decomposing, or containers may be broken or leaking. Never leave a jar with a rusted cap or crack in the cabinet. Spills and vapors are dangerous in themselves and even more so when two or more combine.

If your institution is still storing materials that were once considered the norm but are no longer used, be ruthless. Budget for legal disposal, and don't let "inventory creep" create the same situation again. Recycle your old catalogs and papers. Model green behavior and include your students in the process.

4

Simplify and Reduce

Although college students may look sophisticated and savvy, it's easy to forget they need to explore simple phenomena. Experiences with familiar products can be memorable and have more practical application for your community college students than mysterious alchemy. It is also easier to maintain a safe, organized inventory with small amounts of less hazardous materials than to manage loaded shelves and cluttered storerooms.

THE SAVVY SCIENCE INSTRUCTOR

Great news: The science department was finally going to get needed renovations and all new labs.

Bad news: Everyone had to pack up his or her gear and move to portable classrooms at the far end of the campus for the next year.

Good news: The temporary facilities were better than the rooms that were being abandoned. Everyone would have sinks with hot and cold water, safety showers and eyewashes, and air circulation according to code. Even the storage units in the portables were better. Because all the labs would be clustered together, the chemical and equipment storage rooms would be near the classrooms.

Bad news: The storage space in classrooms and storage rooms was much smaller—only a third of the combined capacity in the old building.

Tough love solution: Pitch in and pitch out. Everyone in the department had to do it.

- Old papers, clippings, magazines, posters, bulletins, 10-year-old science supply catalogs—all were tossed. File cabinets were emptied. The favorite old tests and lecture notes were scanned and saved on CD-ROM.

(cont. on next page)

(cont. from preceding page)

- Broken parts of anatomical models, old engine parts, wiring harnesses from a demo long since abandoned, and a couple dozen power cords were beyond repair and were sent to the recycling bins, along with many more unidentifiable items.
- Old logs, branches, and bracket fungus specimens filled two supply closets. All were destined for the composter.
- Several bags of moldy animal bedding and a 30-gallon trash can filled with bug-infested birdseed were found in a closet at the back of the bio storeroom. They needed decontamination before disposal.
- Buckets contained already-dissected sheep eyes, beef hearts, and fetal pigs. Because they were immersed in formaldehyde, they had to be treated as hazardous waste.
- A drawer full of very old worn-out batteries was found in the physics lab. Several dozen mercury-filled thermometers and weather instruments were gathered from almost every lab. These items were hazardous waste.
- No one expected to find the 500-gram bottle of sodium cyanide and half-exposed sticks of yellow phosphorous—disasters waiting to happen.
- Then came the finds from cabinets in the Earth science, physics, and bio labs: petroleum ether, carbon tetrachloride, lead solder, and concentrated nitric acid turned deep orange in color. Almost every instructor found bottles with labels so damaged it was impossible to identify the contents.
- An emergency contract had to be issued to a hazardous waste removal firm. They came in with chemical suits, masks, and barrels. The final bill: more than $12,000.

It took most of the first month of summer vacation to finish sorting through all that had been saved. Even though the instructors were paid, it was hard work, physically and emotionally. All agreed they had been living on the edge of disaster and vowed never to let it happen again.

4

Connections

- American Chemical Society "Less is Better"
  http://membership.acs.org/c/ccs/pubs/less_is_better.pdf
- Cornell University Environmental Health
  www.ehs.cornell.edu
- U.S. Environmental Protection Agency is responsible for hazardous waste
  www.epa.gov/ebtpages/wastes.html
- Vermont Safety Information Resources, Inc. (SIRI)
  www.siri.org
- Flinn Chemical and Biological Catalog Reference Manual
  www.flinnsci.com

Lively Science

Living Organisms and More

Whether it's called Introductory Biology, Natural History, or Environmental Issues, a life science course is often the first contact a community college student has with the science department. It's also often the only contact nonscience majors have in college with the scientific way of knowing. Almost all of today's students examine ecosystems, mitochondria, and DNA, applying concepts and laboratory skills at macroscopic and microscopic levels. Many courses are introductory to medical careers and also encompass biochemistry, biophysics, biotechnology, biostatistics, and more. So the biology department needs to safely manage chemicals, electronic and optical equipment, and perhaps most complex of all—living organisms.

5

Where Have All the Flowers Gone?

Most life science laboratories today are filled with chemicals and electronics, but at the heart of it all, biology is still the science of life. Observing and studying living organisms is critical to a good, strong program. Amidst all the chemicals, new equipment, and technology, living organisms need to be maintained in a safe and educationally sound manner. This requires serious effort, so remembering just why it is worth all the extra work is important:

- Properly maintaining living things supports a respect for life at all levels of complexity.
- Observing the movement and behavior of living organisms allows students to relate structure with function, stimulus with response, and attain a better understanding of complex biological functions.
- Learning how to maintain healthy living biosystems provides practical experience and adds to understanding of ecological concepts.
- Recognizing the requirements for keeping living things healthy can become a springboard for discussion about healthy personal choices.
- Asking questions about living things provides an opportunity to explore complex concepts that cross academic disciplines.

5

Keeping live organisms in a college science department requires not only knowledge and preparation but also the proper equipment and space. As with selecting pets, selecting classroom organisms requires a realistic appraisal of your circumstances. You must check governing regulations and institutional policy for restrictions and reporting requirements. Then ask yourself: How much space do I have? How much time can I give? How much control do I have? How will plants and animals be maintained during weekends and vacations? Does an organism require a long-term commitment, perhaps many years? Am I willing to commit to a multiyear responsibility? Is the space I will use for cages or tanks shared with other instructors? Is it used by sections that are scheduled during evenings or weekends? Do some students or staff have special needs or allergies that preclude certain types of organisms? How secure is the space—is it used for nonscience functions? Is keeping living organisms the best use of the space that they will occupy?

Resist the temptation to make headlines with your department menagerie. Some animals, such as larger mammals and exotic species, may represent serious safety challenges with relatively low educational payoff. Rescuing injured or stranded animals may seem heroic but may be both unsafe and illegal. The table on p. 73 can help you make informed choices.

YEAR-ROUND ALLERGY SEASON

For a variety of reasons, allergies and sensitivity to environmental contaminants are becoming more and more common. Adults can react to organisms and materials that had never been a problem, because sensitivity can build up from multiple exposures or because heightened sensitivity can be induced by exposure to contaminants outside the classroom. Although some sensitivity may be documented in an accommodation request, many students and staff may not even be aware of allergies to the organisms and materials you are considering. For this reason, avoid organisms known to create problems. Be particularly alert to any signs of sensitivity—sneezes, runny noses, itches, rashes, headaches, increased illness, or absences—when you bring any new organism or material to your facilities.

Latex rubber (commonly used in gloves, lab aprons, and tubing) and nutmeats present the risk of serious, potentially fatal, allergies for some individuals. Choose nonlatex alternatives. Use croutons instead of peanuts for calorimetry experiments. For other conditions, you will have to balance the importance of the experience in your program with the potential severity of a reaction from an exposed individual.

Organism	Level of Care	Potential Problems
Plants	Low: need light and water; minimal care during vacations	▸ allergenic mold on plants or in soil; use sterilized potting soil ▸ toxic plants and plant parts
Aquarium	Low: minimal care during vacations; constant temperature needed	▸ microbial infection from fish, unclean aquaria ▸ zoonotic diseases (e.g., salmonellosis)
Crustacea and snails	Moderate: simple foods; minimal care during vacations	▸ bacterial contamination ▸ exotic species may endanger the environment
Insects, butterflies	Moderate: cultures can become moldy; some care necessary during vacations	▸ stings ▸ infestation of classroom or building ▸ escape or release may endanger the environment
Amphibians (tadpoles, frogs, salamanders)	Moderate: maintain reasonable temperatures	▸ microbial infection from unclean aquaria or terraria ▸ loss of colony due to insufficient heat or cooling during vacations ▸ zoonotic diseases (e.g., salmonellosis) ▸ endangered species may not be collected ▸ escape or release of non-native species may endanger the environment
Reptiles (snakes, lizards, turtles)	High: many species require live food; intolerant of cold; require tight security	▸ bites ▸ bacterial, mold, and other contamination from food and water containers ▸ sensitive to temperature change ▸ zoonotic diseases (e.g., salmonellosis ▸ high escape and release risk
Rodents, rabbits, and other mammals	High: cannot be left unattended during vacations; require tight security	▸ allergenic dander ▸ odor and molds from droppings and bedding ▸ bites and scratches ▸ zoonotic diseases (e.g., salmonellosis, lymphatic choriomeningitis (LCMV), tularemia) ▸ high escape and release risk
Birds	High: need constant care	▸ bites and scratches ▸ zoonotic diseases (e.g., psittacosis, cryptococcosis, salmonellosis, avian influenza) ▸ parasitic mites and ticks

5

5

Who Lives Here?

Signage is important to provide critical information for anyone working in or around laboratory organisms. An animal that's likely to bite shouldn't be tolerated at all, but many animals deserve a "*Do Not Touch*" sign even though they have never demonstrated any aggressive behavior. Healthy-looking animals, along with their litter and bedding, can harbor human infectious microbes, so appropriate gloving, gowning, and masking instructions should be posted clearly. Plants and microcultures should also be identified and labeled, especially if they can cause toxic or allergic reactions.

Biota

Take a closer look at some of the species commonly chosen for classroom culture.

Bacteria

Beginning college students should never culture environmental bacteria or human specimens (including their own) in petri dish systems. Infectious antibiotic-resistant *streptococci* and *staphylococci* are found in the environment. Although a person with a normal immune system can withstand the challenge from small amounts of bacteria, exposure to the millions or billions of bacteria in a broken or carelessly handled culture dish can quickly overwhelm the body's immune system and result in serious infection and disease. You could produce a major strep or staph outbreak with disastrous results. As students progress through your program to more advanced courses, their ability to maintain aseptic technique may be more dependable. Even so, human pathogens should never be used if a nonpathogenic substitution can be made. You must ensure that the culture you use and distribute to students is correct and uncontaminated. Be particularly careful to avoid accidental substitution or cross contamination when more than one culture is in use or stored in the same facility. Keep your custodial staff in mind, and never discard cultures without thorough decontamination.

Here are some suggestions for safer microbiology studies for first year or general studies students:

- Use fresh commercial cultures of nonpathogenic bacteria for staining and microbiologic studies.
- For demonstrations of microbial metabolism, consider using yeast or yogurt-forming bacteria. Souring milk or vinegar formation make good lessons that can be measured with a pH meter.
- Test for bacterial presence and activity without multiplying the bacteria.
- Test strips available over the counter from pharmacies, such as urinary tract infection test strips, can be used to detect the by-products of bacterial action rather than the bacteria itself.

- Test for the effects of bacterial action by using an indicator such as bromthymol blue that measures pH changes from carbon dioxide. Many bacteria release carbon dioxide as they respire.
- Instead of culturing soil bacteria, survey the indicators or products of microbial presence in the soil such as pH, color, and texture.
- Measure soil biodiversity (counting types of macro invertebrates) as an indirect index of soil bacterial action.
- If you must culture environmental bacteria, use closed culture systems available from most laboratory supply houses. Don't incubate environmental samples above room temperature; the most common human pathogens thrive at the warmer temperatures.
- Eliminate all pipetting by mouth. Use mechanical substitutes.

A standard tool of microbiology, the autoclave, can be dangerous if not used correctly. Set up the autoclave in a preparation room, and make sure you follow all safety precautions for the equipment. Pressure cookers have been substituted for autoclaves, but pressure cookers can become bombs, releasing superheated steam; we strongly recommend a gainst their use.

SCLINKS. THE WORLD'S A CLICK AWAY
Topic: bacteria
Go to: *www.scilinks.org*
Code: SSCC75A

5

SERRATIA MARCESCENS

The colorful bacterium *Serratia marcescens* was once commonly recommended for life science activities. But it has been found to be pathogenic and should be avoided if possible. It's one of many organisms that we now understand better—and have learned to avoid. Don't assume that the activities or organisms that were once considered safe are still in the "acceptable" column.

Protists

Protists can be obtained in pure cultures or identified from water samples. In either case, it's important to emphasize proper protection and cleanup when handling, because many serious diseases can be caused by protists. (See the section on maintenance of eyewashes, p. 36, for one example.)

SCLINKS. THE WORLD'S A CLICK AWAY
Topic: protista
Go to: *www.scilinks.org*
Code: SSCC75B

When collecting water from the natural environment, be careful about the source. Do not ask students to collect water samples unsupervised or individually because of the twin risks of drowning and infection. Potentially polluted samples, including common ditch and runoff water, require special handling. When in doubt, assume water is polluted and potentially hazardous. If unpolluted pond water is not available, substitute a hay infusion or commercially prepared sample.

Fungi

Topic: Fungi
Go to: *www.scilinks.org*
Code: SSCC76

The kingdom Fungi includes molds, mildews, yeast, and mushrooms. Many are unwanted visitors in classrooms because of their persistence and tendency to cause allergic reactions. The reproductive spores are the most troublesome. If your curriculum calls for the observation of mold growth, keep the cultures in sealed containers when showing them. Immediately after the activity, disinfect samples with a 10% solution of household bleach and follow with proper disposal in a garbage disposal or sealed bag.

Do not let students bring wild mushrooms into the classroom; some are poisonous even to the touch, and some can provoke dangerous allergic reactions. Use grocery or commercial mushroom samples.

Yeasts make excellent subjects for life science experiments. Baker's yeast is inexpensive and available in grocery stores. It is easily cultured in a sterile molasses-water solution.

MOLDS—GUESTS THAT STAY FOREVER

Once they get established, molds and mildew are almost impossible to remove. They lurk in carpeting and mats; animal bedding and litter; germinating seeds, plants, and soils taken directly from the outside; in the deep dark reaches of lockers and storage cabinets; and in heating, ventilation, and air conditioning (HVAC) systems and ductwork. Mold spores are easily airborne, lodging anywhere, and "blooming" years later when conditions are just right. Ridding classrooms and buildings of these persistent allergens can be very expensive.

Rather than culturing and growing molds, engage students in a hunt for existing mold and spores in the environment. Collect mold samples by pressing a piece of cellophane tape onto the eruption. Place the tape sticky side down on a glass slide. The mold is contained and available for microscopic observation without direct contact or spreading of spores. The slide also provides a valuable record of mold species and contamination location.

Conduct an environmental mold/spore survey by using a glass slide coated with a very thin layer of petroleum jelly. Place the slide in the desired location—perhaps at a windowsill or next to a sink—and leave it for a specified period. After the predetermined time has elapsed, place a cover slip on the slide and examine microscopically. In addition to environmental detritus such as shed skin cells, dust, and pieces of thread, there may be spores indicating a mold contamination. This method may be used to sample for airborne pollen as well. If used outdoors, the slide must be protected from rain or storm. A small open-sided shelter can be constructed easily.

Plants

Plants are attractive, easy to maintain, and provide valuable, timely lessons. But caution is still required. Some common plants or parts of plants are toxic if eaten, burned, or even touched. Beware of such common species as alamanda, oleander, hemlock, poinsettias, dieffenbachia, coontie, and holly berries.

Be explicit. Warn your students not to taste any plants or parts of plants in class, at field sites, or even in their home gardens or yards unless the plants have been specifically grown for food. Even then, they should taste only the edible parts. Also warn students not to harvest endangered plants or distribute non-native species.

Invertebrates

For stimulus-response and life-cycle studies, many invertebrates make excellent choices. Because invertebrate metabolism is so different from that of mammals, invertebrates may be less likely to carry diseases that can spread to humans. Populations of fruit flies, mealworms, pill bugs, ants, cockroaches, hermit crabs, and dermestid beetles can be raised in jars or small terraria. *Daphnia* and brine shrimp may be barely visible to the eye, but their heartbeats can be studied under various conditions with a low-power stereo "dissecting" microscope.

SciLinks — THE WORLD'S A CLICK AWAY
Topic: invertebrates
Go to: *www.scilinks.org*
Code: SSCC77

Although invertebrates are small and relatively easy to care for, some are hazardous and should not be collected or cultured. Some can sting or pinch. The sting of bees, wasps, and fire ants can be very painful and, to the hypersensitive, sometimes fatal. Keep in mind that the bedding or stale food from invertebrate cultures can also be a source of bacteria and mold.

Fish

Aquarium fish can be maintained safely with a few commonsense precautions. Like all classroom animals, fish should come from a reputable dealer. Resist the temptation to bring excitement into your lab or classroom in the form of notorious species such as piranhas, large oscar fish, and live sharks—the risk and liability are not justified. Gloves long enough to keep staff and students from water and moisture contact are advised when cleaning or working in an aquarium. Remind students to use appropriate hygiene, including hand washing and proper disposal of filter materials. Never release specimens into a lake, pond, or other natural water body where they can become voracious predators that dominate and threaten native species.

Amphibians

The common sense rules for care of frogs, toads, and newts are similar to those for fish. Thorough hand washing with soap and warm water after handling them is important. A number of species of amphibians (including one large toad in the southern United

States) are poisonous and should not be brought into classrooms. Many species are endangered, so you should not collect from natural habitats.

In years past it was common to order live animals and get diseased specimens. If that happens, make sure that the animals are sacrificed and the organic material thoroughly incinerated before you dispose of it.

Reptiles

Many instructors find small reptiles are convenient classroom organisms because they provide opportunities for observation and can tolerate infrequent feedings. But even small reptiles require special attention to security and other precautions.

Nonvenomous snakes can bite, and some have skin secretions that can cause severe reactions in sensitive persons. Small lizards and geckos are preferable to snakes, unless you have good security and the experience to keep both observers and animals safe. Snakes and lizards seem to attract theft and vandalism more than many other classroom organisms and have an uncanny ability to escape, so take steps to secure these animals.

Large snakes, such as boa constrictors, can pose a strangulation hazard to smaller animals and small children. Iguanas are very difficult to care for and are generally not recommended. Live feeding of reptiles in front of students is prohibited in many jurisdictions.

Birds

Birds bred by legitimate breeders are generally free from zoonotic diseases (diseases that can be transmitted from nonhuman animals to humans), although every year there are a few reported cases of human disease transmitted by birds. At time of writing, the possible transmission of the H5N1 avian influenza from migratory to domestic birds is of particular concern. If you need to maintain birds, investigate the current research and make sure that your specimens are healthy and isolated from wild species.

In many instances, the greater danger may be to the birds than to humans. Many bird species are temperamental and easily disturbed. When subject to the noises and disturbances common to classrooms and labs, they may become ill or self-destructive.

By-products of birds are also potentially hazardous. Duck and goose droppings and unsterilized owl pellets may also carry agents infectious to humans. Buy commercial owl pellets rather than dissecting those you find in the wild. Feathers can carry disease and mites; they must be sterilized before classroom use.

Mammals

Resist the lure of big eyes and a furry body unless you are very sure you have a healthy, calm animal and you are ready for a lot of work. Mammals can be the most difficult and demanding of classroom fauna and also carry many diseases that can be contracted by humans.

Gerbils and white mice are the easiest mammals to maintain, but they reproduce prolifically. Rabbits, hamsters, and guinea pigs are more troublesome. They scratch and bite, pass on zoonotic diseases, and shed a persistent dander to which many people are allergic. Rare, imported, and more exotic species are much more likely to have been obtained by questionable means from questionable sources and have been responsible for serious disease outbreaks.

If you choose to keep mammals, you should study behavioral characteristics such as hours of activity (e.g., nocturnal, diurnal), grouping (e.g., solo, pairs, colonies), and reproductive activity and frequency, and be sure you can support an animal in an environment consistent with its natural habits. In some cases, pairs or colonies of same-sex individuals will work better than mating pairs, but you must follow appropriate procedures to ensure that the animals' natural territoriality does not lead to fighting and injuries when you first introduce and mingle the animals.

Discourage students from bringing you injured animals, roadkill, or household pets. They could pose danger to you and the students. For most wild animals, a special permit is required. If students report a sick animal, call for expert help from the local animal control officer, the humane society, or agencies specializing in rescuing wild animals rather than intervening.

Humane and Legal

Strict federal and state rules govern the use of animals for research. Although maintaining classroom specimens for observation wouldn't normally be considered research, this use might be a gray area at some institutions.

If you use your laboratory animals for any experiments at all, investigate regulations governing their use. Don't do anything on the spur of the moment; discuss contemplated activities (including observations under various conditions) ahead of time with your department head.

Aliens—Escape and Release

Every time you bring an organism into your classroom, you gain a friend for life—and a responsibility. Non-native organisms must never be released into the environment. Nor should your colleagues have to contend with your freedom-loving fruit flies, mice, or reptiles.

Before you start a classroom culture, think about the end of the term first. Ducklings may be returned to a farm, native plants to an old field. But do not plan to release purchased butterflies or second-generation mice into your neighborhood. They can spread disease, upset the balance in the habitat, and cause many other problems.

Exotic species can cause tremendous environmental damage and serious disease. Organisms that are taken from their native environment may carry microorganisms

Call of the Wild

Do not bring wild or feral animals—dead or alive—into your classroom. They may

- be sick, diseased, or hurt
- carry infectious agents, ticks, lice, or other parasites
- be protected by federal, state, and local regulations
- require special licensing
- be much more aggressive than domesticated animals

that are indigenous and innocuous to other species in the native environment but could cause serious disease and death when they cross over to species not normally found in the native environment. Unchecked proliferation can result from lack of natural predators. Nutria that are gradually dismantling Louisiana's bayou country and the iguanas on Florida rocks were once someone's pets. Destructive gypsy moths and aggressive Africanized bees are the unintended consequences of scientific studies. Kudzu, purple loosestrife, and dandelions were originally imported as attractive ornamentals. Instructors should model responsibility by not using or maintaining exotic organisms if it can be avoided.

5

Bacteria and other microcultures can be exotic as well. Some classic laboratory experiments involve purchased bacteria, such as the formation of crown galls with *agrobacterium.* This is not only illegal in some states, but a serious risk to plants in most communities if the experimental species escape to the local environment—including via unincinerated trash. Unless you have excellent training, security, and disposal facilities, the best practice is to work with native species at all times.

Students as Living Laboratories–A Bad Plan

Traditional biology labs have used students as test subjects and as sources of tissue and fluid samples. Biology instructors should reconsider these practices in light of better information about communicable diseases.

- Consider all blood and body fluids to be infectious, and refer to Chapter 10, p. XX, for additional information on Standard (Universal) Precautions. Make sure you have a specific plan for disposal of all biohazard material and that everyone who uses your classroom knows the plan and the specific procedures necessary for disposal of these materials.
- With introductory students, we recommend you not conduct any laboratory work that involves the use of human body fluids (e.g., saliva, urine, blood, sera, semen). Do not, however, avoid topics concerning body fluids. The college beginning biology class can present many opportunities to discuss the transmission of infection via human body fluids; the factual scientific information may be invaluable to students engaging in risky personal behavior. We recommend discussion rather than laboratory activities with human body

SCiLINKS. THE WORLD'S A CLICK AWAY
Topic: viruses
Go to: *www.scilinks.org*
Code: SSCC80

fluids. Be certain that students in advanced classes are fully instructed in Standard (Universal) Precautions before considering activities involving human body fluids.

- Do not collect, type, test, or otherwise use human blood. If your curriculum requires a blood-typing demonstration, use newly purchased kits with synthetic antigens. It is not just using your own or your students' blood that presents a risk. The antigens that were sold in older blood-test kits, which may still be hanging around the lab, were prepared from untested blood purchased "on the street" and could present hepatitis and HIV risks. Nursing and health care students should have very specific preparation before they begin to handle body fluids.
- Some of the newer test strip systems (for glucose and urinary protein) use bioengineered antibodies, and can be reasonably used by students in private or at home, but urine specimens should not be brought into the classroom.
- Use prepared amylase, not saliva, to hydrolyze starch.
- If you use straws or swabs, have a container of disinfectant or specific disposal system available at each work station. Never reuse any material that has been exposed to human body fluids. Have a disposal plan, and instruct students to discard contaminated items immediately in a manner that does not pose risks to the maintenance staff.
- Do not use pins or any other sharp implement for sensory nerve assays. Use coffee stirrers or hair roller picks, and properly dispose of all used items.
- Before beginning any studies that involve exercise or other means of increasing a student's heart rate, temperature, and other vital signs, make sure you inform students of the activity and receive written confirmation that they have been informed and are medically able to conduct the study safely. Provide them with an easy but private method to communicate health problems to you that may preclude their participation or require modifying the activity for them. Never schedule these activities during hot and humid weather, and warn students to stop immediately if they feel any discomfort or abnormal response.
- Before planning any activities that involve measuring students' personal characteristics, such as height, weight, body type, or family history, consider if one or more of your students might be embarrassed as a result. If you cannot protect those students, you must find an alternative activity.

Body Fluid Spill Kit

All labs should have a blood or body fluid spill kit easily and quickly accessible. It may be prepared in-house and should at least include protective gloves, eye protection, and an apron. Other items may be included, such as a disinfectant and cloth or paper towels to lay on the spill. Custodians should have special training and be equipped with major spill equipment and supplies for cleanup of human blood or body fluids. All human body fluid spills must be treated as infectious.

SCILINKS. THE WORLD'S A CLICK AWAY
Topic: AIDS
Go to: *www.scilinks.org*
Code: SSCC81

- Remember that some of your community college students may still be minors under the law and parental permission may be necessary.

HOUSEKEEPING IN BIO LABS

A life science classroom has special cleaning requirements:

- Make sure you always have a supply of liquid soap, paper towels, and hot water for hand washing and hygiene. If you use the newer hand-cleaning gels made of quick-drying alcohol formulations, use them cautiously and remember that they are flammable.
- Following life science activities, clean desks and counters with warm water and soap or a 10% solution of household bleach.
- Remember that stains stain. Use minimal quantities and determine, in advance, how to clean and neutralize spills. Keep all biological stains secure so students aren't tempted to use them for tattoos.
- Check the toxicity and hazards of each biological stain before using it. Many are suspected of being carcinogens or teratogens. Avoid the use of hematoxylin, methyl red, or methyl orange (known carcinogens).
- Designate a disposal container for biological waste. Make sure that other instructors and your custodial crew know and follow your procedures.
- Using certain disinfectants and pesticides in buildings with public access may be covered by federal, state, or local regulations. Be sure you know and conform to the latest rulings. In general, only trained personnel may handle certain regulated disinfectants and pesticides and occupants must be notified in advance of their use.
- Use heat-resistant gloves or mitts for handling hot labware.
- Have nonlatex gloves available for use in cleaning spills and stains and handling preserved specimens.
- Conjunctivitis and other infectious diseases may be spread from one user to another via contamination of safety eyewear and optical instrument eyepieces. Refer to Chapter 10, p. 162, for more information.

Partner With Your Support Staff

Respect and protect the custodial staff. They will be in the room working when you and the students are gone for the day. Have you informed them of your plans to keep living organisms? Do they know what is in the room and any special considerations that may be required? Are they aware of any particularly difficult, potentially dangerous, or prone-to-escape organisms? Are they aware of your disposal methods and what they may encounter in the trash? Is special handling required for some waste material?

Develop and maintain a positive relationship with the night crew. They are a vital part of the education team. When they are included and understand your classroom goals and responsibilities, they can more effectively support your program. An apparently simple need for adequate paper towels will no longer be a struggle. Should you forget to water a plant, they will have taken care of it, because they are included as a part of the team.

The Dissection Dilemma

During the past decade, dissection activities have become controversial. Many traditional activities are simply exercises in rote memory, and taking dissection specimens from the wild depletes natural populations. There are safety hazards, too. Sharp instruments pose cut and puncture risks. Dissection preservatives are toxic and often allergenic. On the other hand, proponents of the dissection experience argue that students can learn a lot with carefully supervised dissections structured around inquiry questions rather than just finding and naming parts.

Don't include dissection just because you used to do it. Only use it when it will help students fulfill specific, significant learning goals. If you do decide to dissect, precautions are in order. These precautions apply to all students, but are especially important in introductory classes. Treat all preservatives and preserved specimens as toxic and hazardous. Even if you cannot smell a preservative, it may be toxic. Don't purchase any company's "secret formula" without obtaining and retaining a material safety data sheet (MSDS). (See Chapter 4, p. 57, for a sample sheet.)

Do not allow students access to stock containers containing preservative fluids. Use the lowest toxicity preservatives possible, and review the MSDS for hazards, precautions, and disposal procedures. Do not keep specimens and solutions from year to year, because molds can grow even on preserved specimens. Chemical splash and impact-resistant safety goggles, protective gloves, and an apron must be worn when dissecting. (See Chapter 10, pp. 161–163, for more information about eye protection.) Special precautions must be taken for disposal of both the chemicals and the biological material.

Some instructors have substituted dissection of fresh materials, such as chicken wings and beef hearts, obtained from butchers or grocers. In doing so, remember that unpreserved materials are also common sources of *E. coli* and *salmonella*. You will need soap, hot water, and an adequate supply of paper towels for hand washing; a disinfectant for tools and countertops; and a plan for disposal of the organic waste. Never dissect wild organisms of any kind, or organisms that may have been exposed to wild species.

SCILINKS THE WORLD'S A CLICK AWAY
Topic: salmonella
Go to: *www.scilinks.org*
Code: SSCC84

Students or instructors have prepared skeletons by using dermestid beetles and other processes. But roadkill and fallen animals should not be brought into the lab. If you prepare skeletons, use laboratory-raised or butcher's specimens, protect yourself with gloves, and use bleach to pretreat specimens.

Beyond Your Classroom Door

Your obligation to safety doesn't stop at your classroom door. Responsibility extends to other members of the school community and to the environment around you.

- You need an animal care and security plan for nights, weekends, vacations, or when there is a substitute instructor.
- An active life science program results in chemical, biological, and other hazardous wastes. Ensure safety along the entire waste disposal stream.
- Be especially cognizant of your obligations to the custodial and maintenance staff.

Biological Chemicals

Life science activities involve as many toxic and hazardous chemicals as chemistry activities. So biology laboratories require adjacent storage and prep areas outfitted similar to chemistry laboratories.

Many biological chemicals are even more hazardous than those used in first-year chemistry. They can be insidious, because organic chemicals that are carcinogens or teratogens may not cause immediate health effects. They may also be taken for granted because only a tiny amount would normally be used. Consider the stain hematoxylin. A professional cytotechnologist might safely use small amounts in solution for an entire batch of slides and handle it without direct contact. But, in a classroom, we have to assume students *might* have a significant skin or respiratory exposure to this known teratogen and that a student *might* be pregnant without our knowing. So our standards of use in the classroom must be much stricter than in professional laboratories.

Many genetic engineering and DNA lab activities use stains even more toxic than those used in standard beginning biology lessons. Because these laboratories are often restricted to advanced students, the stains can be better controlled than they would be in a first course. However, biological storerooms may be shared with other instructors, and you may not be aware of a pregnant colleague or student who could be exposed, so be very cautious with known teratogens. Opt for less toxic nuclear stains such as methylene blue over more dangerous ones such as ethidium bromide or toluidine blue.

Nondenatured ethanol may be used for a variety of legitimate purposes in biology laboratories. However, besides being highly flammable, the item may be a temptation for illegal recreational use. Stocks of nondenatured ethanol must be safely stored in a secure location with access and use closely supervised.

Don't take anything for granted with chemicals or with students. Let students know that you will provide a complete description of every lab well before the activity is scheduled, and tell them to read the cautions and contact you privately if they have any concerns. Research the most recent MSDS information for every chemical you use, and employ appropriate disposal procedures for anything that is questionable in your storeroom.

SCILINKS. THE WORLD'S A CLICK AWAY
Topic: hazardous waste
Go to: *www.scilinks.org*
Code: SSCC85

5

CLEANING HOUSE

Many chemicals once commonly used for biology activities are now known to be unnecessarily toxic or hazardous. Some have been banned outright. It is our recommendation that the following materials be eliminated from labs and substitutes made for the chemicals or activities. Removal must be made by a bonded (in some jurisdictions, licensed) hazardous waste removal service.

- Toluene and benzene—used for embedding samples.
- Ethers and chloroform—used for chromatography or for anesthetizing insects.
- Heavy metals—such as elemental mercury or chromium compounds commonly found in stains and metallic salts.
- Colchicine—once used for mitosis slides.
- Chloroform—once commonly used for insect specimen preparation.
- Carbon tetrachloride—a once commonly used solvent.
- Hematoxylin, methyl red, methyl orange—biological stains.
- Benedict's solution—used for sugar testing, splashes dangerously when heated. Test strips are cheaper and safer.

(cont. on next page)

(cont. from preceding page)

Less hazardous methods and materials are constantly developed and made available. Keep up with these developments and substitute safer methods and materials as they become available. For other reagents, refer to storage rules in Chapter 4, pp. 56, and recommendations in Appendix A, p. 205. Make sure the storeroom is secure, and that sections for combustibles, corrosives, and organics are separate.

You may spend more money, but you are better off ordering chemicals in the strength you need rather than in more concentrated forms for later dilution. Order only what you need for one year—programs may change and chemicals may decompose or become contaminated.

5

PTC

PTC (phenothiocarbamide or 1-phenyl-2-thiorea) "tasting" has been used extensively to demonstrate Mendelian genetics. It is now known that PTC tasting is not a single-gene dominant/recessive trait, so the model is incorrect. Additionally the PTC chemical is toxic (frequently sold as rat poison) and should not be used for any tasting experiments. Find out more at *http://physchem.ox.ac.uk/MSDS/PH/1-phenyl-2-thiourea.html.*

Formaldehyde

Formaldehyde solution—also called formalin—has been replaced largely by newer, less odorous and less toxic preservatives with a variety of trade names. However, most of these new formulations still contain formaldehyde, albeit in lesser concentrations and mixed with other ingredients meant to reduce odor and be less irritating. A prudent approach is to treat all specimen preservatives as though they contain formaldehyde, particularly if you or your students are sensitive to materials of this type. The Flinn Scientific Catalog/Reference Manual and Safety in School Science Laboratories contain good discussions of the issue.

THE SAVVY SCIENCE INSTRUCTOR

Several of the faculty members at Home Town CC are adjuncts with day jobs as professional scientists. Mr. A instructed two evening sections of Natural History 101 but worked for the state fish and game department as an ichthyologist by day.

(cont. on next page)

(cont. from preceding page)

Deciding to enhance his lab exercises with some real-world experience, Mr. A brought in a half dozen specimens that had been collected following a recent fish kill. Between the Thursday section and the Monday section, because he did not have his own refrigerator, he stowed the specimens in the nearest one—in the faculty lounge. Unlabeled, Mr. A's "treat" was unwrapped to the less-than-welcome surprise of Ms. T, the full-timer whose turn it was to perform the usual Friday cleanup of the lounge refrigerator.

Difficult as it was to schedule, the department chair had to put together an all-department meeting and begin work on some long-needed reforms and policies to ensure safety and faculty harmony. A subcommittee of two full-time and two adjuncts negotiated a strictly enforced set of rules for use of common areas—classrooms, labs, storage and prep areas, and the lounge. As limited as space was, a private prep and storage area was assigned to each adjunct faculty member. And a new laboratory refrigerator was ordered for the biology preparation room to be used strictly for specimens and nonvolatile reagents. The faculty lounge refrigerator was designated off-limits for anything but lunches and snacks.

5

Connections

- American Society for Microbiology
  www.microbe.org
- Biosafety in the Laboratory: Prudent Practices for Handling and Disposal of Infectious Materials
  http://books.nap.edu/catalog/1197.html
- Biosafety in Microbiological and Biomedical Laboratories (BMBL) 4th Edition (n.b., This volume issued jointly by the CDC and NIH was being updated at press time. Search for the 5th edition.)
  www.cdc.gov/od/ohs/biosfty/bmbl4/bmbl4toc.htm

(cont. on next page)

5

(cont. from preceding page)

- Centers for Disease Control (CDC)
  www.cdc.gov/healthypets/
- Flinn Scientific Catalog/Reference Manual. 2002.
  Batavia, IL: Flinn Scientific, Inc.
  www.flinnsci.com
- Guidelines for Ethical Conduct in the Care and Use of Animals
  www.apa.org/science/anguide.html
- Hand washing. PMC ISLA Health System—Guam
  www.pmcguam.com/resource/handwash.htm
- Institute for Laboratory Animal Research
  http://dels.nas.edu/ilar_n/ilarhome/reports.shtml
- Institutional Animal Care and Use Committees (links to extensive information)
  www.IACUC.org
- MSDS for Infectious Substances-Health Canada
  www.phac-aspc.gc.ca/msds-ftss/index.html
- National Research Council Guide for the Care and Use of Laboratory Animals
  http://books.nap.edu/catalog/11138.html
  http://books.nap.edu/catalog/5140.html
  Precollege animal care brochure
  http://dels.nas.edu/ilar_n/ilarhome/Principles_Guidelines.pdf
- PTC Tasting physchem
  http://physchem.ox.ac.uk/MSDS/PH/1-phenyl-2-thiourea.html

Modern Alchemy

Safer Teaching With Chemistry

Ever since Democritus gave his first "midterm" on atomic theory, instructors have been saying: "Students just aren't what they used to be." Since then, successful instructors have recognized the truth of his lament and adapted their methods and materials to the students they get—not the ones they wish they had.

6

Chemistry for All Students

High school chemistry courses used to be filled almost exclusively by students contemplating science majors in four-year colleges. Today, many more students take chemistry at the secondary level than a decade ago, but community college instructors must not assume that these students come with the same knowledge and laboratory skills as students arriving even five years ago.

Although more students have been encouraged to take high school chemistry, their teachers have had to modify curriculums to accommodate students with a much broader range of interests and abilities, including students with significant learning disabilities and English language deficits. Courses in "applied" physical sciences have been developed to blend vocational and scientific skills. Additionally, many secondary laboratory activities may have been reduced due to lack of adequate space and equipment, emphasis on standardized testing, and concern for liability in case of accident.

If you can count on anything, it is that you will have to accommodate more diverse needs among your students and less consistency in what they know and can do. As a result, you need to modify your instruction to address the wide range of student interests and abilities and provide all students with the knowledge and the skills to work more safely in science laboratories in class and career.

Looks Can Fool You—Never Assume!

Students may not be the only ones who bring misconceptions to their college chemistry classes. As their instructor, you may be relying on unwarranted assumptions about your students' understandings, abilities, and experiences. With today's lifestyle emphasizing speed and reducing concentration, your most sophisticated-looking students may never have looked at a burning candle carefully enough to notice that the solid wax melts before it burns. Do not allow your students' superior ability to text message and cruise the internet from their cell phones fool you into thinking they know what the inside of an ordinary flashlight battery looks like. On the other hand, that elderly lady at the back of the room who never says anything in class may be better able to explain the structure and function of her car battery than you can.

Additionally, don't assume that all or any of your students have such fundamental skills as being able to interpolate or extrapolate for the unmarked intervals on a measuring instrument, that they recognize the need to clean glassware before and between uses, or that they know that they must never return reagents to the supply bottle. Make sure your students can identify variables in an experimental system and that they know a controlled experiment should have only one variable. Be prepared to teach observation, measurement, and record-keeping skills and procedures.

Today more than ever, college chemistry instructors need to spend a significant amount of time teaching students to observe more carefully and concentrate for longer periods of time than our modern lifestyle normally allows. Students, young and old, have come to have shorter and shorter attention spans and to expect loud, flashy messages that catch their attention, perhaps because they are bombarded with fast-paced media presentations. Natural phenomena may be far more subtle. Students need quiet time to think and review what they have observed—to analyze what they have seen and compare it to their preconceptions. They need to measure carefully and consider both the precision and reliability of the instrument readouts that make experimentation seem so much easier than in the good old days.

Bigger and Louder Is Not Necessarily Better

One of the most dangerous misconceptions a chemistry instructor can have is that it takes an explosive demonstration to motivate students. An explosion more likely creates excitement—from having been placed in a dangerous situation—rather than motivation. Think about it for a moment. If you need to put a safety shield in front of your student audience to ensure that they will be protected, you must be creating a situation with a known hazard—and, by the way, you are standing closer to the hazard, without the benefit of the safety shield, than your students.

The very excitement of the explosion is more likely to stop any critical thinking that students might have been engaged in than to provoke more analytical discussion. And besides, how spectacular can your demos be when compared with what is now routinely produced by special effects? When you were a student, you may have been unforgettably impressed by a loud, explosive demonstration of the reactivity of sodium in water. But recognize how such a demo pales when compared to the light-saber duels in the latest *Star Wars* episode and the explosions released in the battles for Middle Earth.

It's true that the unanticipated creates teachable moments, but we suggest that you startle your students and gain their attention with discrepant events rather than noisy dangerous ones. Use only those demonstrations and experiments that you have tested in advance. Be sure you understand any hazards or precautions for any chemical that you may use. Consider whether your students have the background fundamentals to understand the concepts illustrated, and make sure that you have an MSDS for *every* chemical that you use.

> **Motivation From Unanticipated Results**
>
> ...or when was the last time you noticed a discrepant event such as:
>
> - Archimedes' buoyancy—Eureka!
> - Fleming's discovery of the antibacterial activity of penicillin
> - Marie Curie's observations on the behaviors of radium
> - Röentgen's discovery of an X-ray image

Think Small

Even when you are using familiar substances, accidents can occur. Even in the most disciplined lab, the urge to sneak a taste of something or snatch a bit of chemical to try something at home may be overwhelming. Using the smallest possible quantity of a reagent reduces the magnitude of harm the chemicals can cause. Think drops rather than full test tubes or beakers. With solids, think in granules and pea-sized proportions.

Many excellent microscale chemistry activities have been developed for the college level. Observing a phenomenon on a small scale requires students to pay close attention to small changes and focus on detail. Set students up to discover the excitement of seeing a hypothesis confirmed, rather than seeing an experiment self-destruct.

Microscale experiments appear in many textbooks and on the internet. At a minimum, you need two types of equipment: microchemistry test (well) plates and disposable pipettes. The well plates come in several forms. Those with larger wells are easier to display on an overhead or with a video cam but spill or splash more easily. Wash all labware carefully, because at microscale, even minute quantities of contaminants can affect results. To minimize the chance of contamination, it's a good idea to keep a separate set of test plates for each type of experiment.

We've heard some instructors say: "Microscale doesn't work because my students don't notice what I want them to see." Consider these two possibilities: First, observation is a skill, and teaching students to observe well may need to precede other lessons. Second, it's possible that because of misconceptions, students are looking for results that are surprisingly different from what you expect them to notice. Listen carefully to identify students' unconscious assumptions when they are working in the lab.

In instances when microchemistry isn't practical, you can still scale down the quantities you use to reduce hazards. By keeping the stock reagent bottle out of the

6

Student Assistants

Almost every community college has a laboratory coordinator and paid or credited student assistants. The lab coordinator is likely to have specialized training in physical science and normally trains the student assistants. But, remember, you are ultimately responsible for everything that happens related to your course. Even the best-trained students should be permitted to enter chemical storage rooms only to obtain specific supplies for specific purposes. They should not be permitted to work in preparation rooms unless a supervisor is available to answer questions and to take care of emergencies. The ultimate responsibility is always yours.

classroom and the quantity present for distribution very small, the consequences of human error can be minimized.

Guard Your Stock

The stock supply of any chemical, even those you think are harmless, should remain in a locked cabinet or storeroom. (Refer to Chapter 4, p. 56, for more detailed information on chemical storage.)

Security is especially important with hazardous items such as corrosives, reactives, flammables, and toxics and materials that have been reported in the media as ingredients for explosives. Theft of reagents is unfortunately all too common and continues to increase as more illicit operations are set up to take advantage of "recipes" found too easily on the internet.

Before class, determine the quantities you need, bring only that quantity to the open laboratory, and make sure that all containers are correctly labeled. Clean plastic trays or storage tubs are handy for organizing and transporting supplies for specific lab exercises.

Preparations should not be taking place in the chemical storeroom where an accident can result in disastrous combinations of the reagents stored there. Prep rooms should be separate from the chemical storage room and in an area equipped with the proper safety gear.

Some of the most serious lab accidents have occurred when the stock supply of a flammable liquid such as alcohol was inadvertently placed near a heat or spark source. *Stock containers of alcohol and any other flammables should never be in the open lab.*

Moderation and organization are keys to safety. The guidelines in this chapter and "Chemicals to Go" listed in Appendix A, p. 205, are designed with the introductory student in mind. If your institution has an advanced program such as organic chemistry, it's best to keep chemicals for those classes in a separate, secured storeroom or separately secured storage cabinet within the general chemical storeroom. Access to chemicals for advanced classes should be limited to those who need them and to those assistants specifically trained to handle those reagents properly.

Do not keep reagents that are no longer needed. Be prepared to purchase new chemical supplies rather than save chemicals that might be used again in the future. Old, unused reagents tend to accumulate dangerously, especially if faculty members who originally ordered them are no longer on staff. At the community college level, the institution should have a licensed, bonded hazardous waste removal firm under contract, and instructors and lab personnel should dispose of outdated, unused, and otherwise hazardous reagents at the end of every term or at least once every year.

Bad Combos

Many chemicals can cause violent reactions or create highly toxic gases when combined. That's one reason why a chemical storeroom should contain only the materials necessary for the current curriculum in the least amount possible. *To avoid inadvertent deadly combinations, storage of chemicals should be based on reagent characteristics and not alphabetical order.* Store incompatible chemicals separately and in appropriate containers. Store acids and bases separately and strong oxidizing agents apart from strong reducing agents. When adding new reagents, check reliable websites for the latest information on incompatible chemicals. Two that we have found useful are:

- *www.pp.okstate.edu/ehs/HAZMAT/LABMAN/Appendix-B.htm*
- *http://ptcl.chem.ox.ac.uk/MSDS/incompatibles.html.*

Oldies but Not Goodies

Many of the chemistry demonstrations and materials found in old books are now known to be hazardous. Refer to Appendix A for a partial list of chemicals that do not belong in an introductory college program. Review all your old favorites by checking the MSDS and a good, current website reference to see if other items should be abandoned as well. You may find that you need to request hazardous waste removal to get rid of your predecessors' leftovers.

Be alert to cloudy films or powders collecting on the outside of glass bottles or the insides of cabinets and other surfaces. These are usually signs of leaks or fumes from chemical containers. Cracked or corroded caps on containers indicate reactivity or contamination; the contents should be removed carefully since purity cannot be guaranteed.

Chemicals cannot simply be thrown away. Disposal is a major issue for laboratories. Many of the materials cost far more to get rid of than to purchase. Hazardous waste disposal is governed by requirements of the U.S. Environmental Protection Agency Resource Conservation and Recovery Act and should be conducted only in conjunction with reputable, licensed firms. If you check into the cost of disposal, it will make microchemistry all the more appealing. (Refer to Chapter 4 and Appendix A for more information on chemical storage and hazardous waste disposal.)

No Wiggle Room

Bumping and jostling are frequent causes of spills and accidents during chemistry activities. A way to minimize these risks is to provide as much clear work space as possible. This means you should ask students to leave all but the essentials at home or in their cars when they come to class. Tell them in advance when specific items such as texts are needed and ask that no extra books and materials be brought to lab classes. If the parking lot is too far away for students to retrieve their possessions between classes, arrange a space in some part of your room, away from work surfaces, where briefcases, backpacks, and other paraphernalia can be deposited during lab activities. Select this area carefully so that exits or paths are not obstructed or blocked in an emergency. Explain how sudden or extraneous motion can be especially dangerous during chemistry activities.

In general, it is better for students to stand while doing chemistry experiments to avoid the possibility of spills getting onto laps. Remove chairs and stools from the lab area so students have room to step back or make a quick exit. Sitting should be permitted only for physically disabled students and then only at benches designed for handicapped access. If you are stuck with those antique bench designs with the high shelves between students, request that they be removed as soon as possible and forbid students from placing items on them.

Tools of the Trade

In keeping with the idea that simpler and smaller are better, look over your inventory of science equipment. Some of what you own may be more hazardous and less useful than newer counterparts. If you have students preparing for technical positions, it is especially important that you provide them with experience on up-to-date equipment and technology.

Mercury-filled instruments: If you still have barometers, thermometers, or other measuring instruments made with mercury (the column is silver in color), you probably need to arrange for safe disposal immediately. The danger—and cleanup costs—posed by spilled mercury from broken instruments is serious and unacceptable. Some instructors prefer mercury thermometers for advanced work, but on closer examination, this isn't a valid reason to risk the mercury hazard. These instruments have electronic counterparts that are more precise and no longer prohibitively expensive.

Instruments for measuring mass: Keep a few double-pan or triple-beam balances to help students review the concept of mass—they are great for your history of science lecture. But once the idea is well established, you will find that using electronic balances reduces the time required to take measurements and can produce more precise data. Lock up all your balances when you are not using them. If you plan to leave balances on counters or desktops when a room is not supervised, have them secured to the casework with bolts or chains. Balances are among the most commonly stolen items from science facilities because of their value in illicit drug activity.

General-use containers and instruments to measure volume (e.g., beakers and graduated cylinders): Use only plastic or borosylicate glassware. If using plastic, review manufacturer's specifications to be certain that the material is compatible with the use you intend.

Discard any labware that is cracked or chipped. Even glassware that looks safe can fracture, so tape glassware when it will be under pressure or evacuated and ensure that everyone is wearing safety eye protection that provides impact resistance and splash protection. (See Chapter 10, pp. 161–162.)

Never use any laboratory containers for food or drink.

Sharp instruments: Wear eye protection when using cutting tools such as utility knives or scalpels and glass instruments. You must have clearly labeled sharps disposal containers to protect personnel who handle the trash. (See Chapter 10, p. 157, for additional discussion on sharps.)

Tubing and connectors: Whenever possible, substitute plastic tubing for glass tubing and latex rubber tubing. If you use glass tubing, make sure it is all fire polished.

Students should not cut, bend, or fire polish glass, or insert glass tubing into corks or stoppers until you have given them very specific training. Exercise extreme care if you engage in this work. Cutting and inserting glass involve serious risks from breaking glass and can cause permanent injury to hands and eyes. Bending or fire polishing glass adds the danger of serious burns, especially when hot glass does not look hot. Wear impact-resistant safety glasses and gloves. Use heat-resistant mats when you set down hot glass, and always protect your hands with something like a thick towel if you insert glass tubing. There are some tools that assist with glass tube insertion, but use them with careful attention.

Asbestos-containing products: In an older storeroom, you might find asbestos mats or mitts for handling hot glass or protecting work surfaces. Asbestos-containing products should be replaced. They cannot be thrown into the regular trash but must be turned over to a certified asbestos disposer.

Spark igniters: These should be checked before use and secured after each lab.

Spectrometers and other electric and electronic equipment: Check the integrity of wiring and keep equipment away from water. All receptacles should have GFCI (ground fault circuit interrupter) protection.

Electrophoresis apparatus: In addition to the hazards associated with electrical equipment, higher voltages may pose additional risk.

Calorimeters and gas generators: Make sure they have properly functioning pressure releases.

Heat sources: Bunsen burners are standard for some college chemistry experiments but may not be needed as often as they once were. When a lab-certified hot plate can be used, do so but remind students that the surfaces will remain hot long after the hot plates are turned off. Your gas supply should have an emergency shutoff accessible for use in an emergency *and* a locked shutoff for times when the burners are not needed. Gas should not be available to students at times when they are not specifically required to use it. Personally check all burners before use. *Never use alcohol burners.* (See Chapter 3, p. 42–43, for additional explanation.)

Gas cylinders: Once familiar pieces of science equipment, compressed gas cylinders are used less frequently these days. Don't store compressed oxygen or hydrogen unless there is a very specific need to do so—and then in a secured location separate from the introductory chemistry materials. If there is a valid reason to have compressed gases in your lab or prep area, observe the following precautions:

- Obtain and refill your cylinders through licensed, reputable dealers.
- Examine valves and handle them carefully. Don't lift the cylinder by the valve, and make sure the valve is closed firmly when not in use.
- Keep the labels intact and clear.
- Store cylinders in an upright position secured with a chain or clamp.
- Store cylinders in a cool place, never near heat.
- Protect cylinders from bumping and tipping.
- Keep the protective cap in place except when the cylinder is in use.

Gear Up for Safety

In the new freedom of a college environment, students often assume that anything goes with respect to dress. Don't be intimidated by that attitude. It is legal and reasonable to demand appropriate dress in a college chemistry class—and it's great training for the world of work. Have students research and review Occupational Safety and Health Administration (OSHA) requirements for laboratory and workplace safety. Rather than simply banning certain types of clothing, it may be useful to involve students in assessing the hazards. In your introductory safety lesson, you might include a demonstration of the flammability of various fabrics, hair gels, and accessories to help establish laboratory dress rules and demonstrate the reasons for the rules. Product safety and workplace safety are important issues in commerce, so these experiences may suggest some interesting and attractive career options.

Safety eyewear: Every chemistry experiment requires safety eyewear. In addition to the potential for chemicals entering the eyes, you must protect against the possibility of projectiles resulting from a dropped or shattered container and materials sent flying by an unanticipated release of pressure. If there is any doubt, have everyone put on safety goggles—this includes all room occupants. See Chapter 10, p. 161, "All Eyes on Science," for a detailed explanation of eye protection.

The most straightforward way to handle safety eyewear is to require each student to buy a pair of safety goggles. Make sure that what students buy is certified for both splash and impact resistance. If individual purchase is not possible or violates an institutional policy, safety eyewear must be sanitized between uses. This can be done by hanging the eyewear in a sanitizer equipped with ultraviolet (UV) lights. However, if students in the next class must reuse the eyewear immediately, there may not be enough time for disinfection. Goggles piled haphazardly into UV sterilization cabinets might not get sanitized at all. An alternative procedure is for departing students to drop their goggles into a sink filled with antibacterial dishwashing solution. Entering students can thoroughly rinse and dry the goggles and wear them immediately afterward. If you use this method, make sure straps are plastic rather than cloth. Have plenty of soft clean towels available, because rough paper towels can scratch the plastic lenses.

Equipment Inventory

Keep an up-to-date list of equipment. If possible, store the equipment so that a missing piece is noticed immediately because a space on the shelf or in the storage area is vacant. Periodically review all your equipment for condition and usefulness.

Pay particular attention to AC-powered instruments and equipment. Is the power cord complete, uncut, and unfrayed? Is there an intact three-prong plug on the end of the cord? If the plug or cord is damaged, have it replaced—not just patched but also professionally repaired. An improper money-saving fix could be very expensive should it be found to have contributed to an accident. Clear the storeroom of broken and unused equipment. Broken equipment set aside for "someday" is a lost resource. Fix it or dump it, but resist the temptation to just keep it.

Face shields: If an experiment requires a face shield, it probably shouldn't be done at the introductory college level. However, you may need to use a face shield in addition to your safety goggles to protect the rest of your face and throat when you prepare materials for classes. Use the face shield when diluting concentrated acids, although we recommend purchasing prediluted solutions. If the experiment is absolutely required in the curriculum, consider using microscale quantities and a video cam to reduce risks.

Aprons: A waterproof flame-retardant lab apron should be used whenever there is a possibility of spills, splash, or flame. When in use, the apron ties should be securely fastened. Worn correctly, the apron can also be used to hold back loose-fitting clothing that poses hazards from knocking something over or catching on fire. Loose-fitting sleeves should be rolled up, fastened with elastic bands, or both. Avoid latex aprons and gloves because of the possibility of severe allergic reaction in sensitive individuals.

Chemistry Students Dress for Success

- Use impact-resistant chemical splash safety goggles throughout every lab.
- Fasten hair securely behind the shoulders.
- Do not wear loose-fitting clothing or headgear.
- Minimize unprotected skin (e.g., bare midriffs).
- Use lab aprons and rubber bands to hold back dangling items and loose clothing.
- Always wear shoes. Discourage the use of shoes, such as sandals, flip-flops, and platforms, that provide insufficient covering, stability, and support.
- Remove or tape over dangling jewelry.
- Use appropriate gloves when needed.

Gloves: Use heat-resistant gloves or mitts for handling hot labware. Have nonlatex gloves available for use in cleaning spills and stains.

Tongs: Tongs may be useful tools for the instructor handling hot labware, but students, who do not have as much laboratory experience, may be less adept at managing them and spills often occur. You may be able to avoid having students use tongs by preparing materials ahead of time and letting them cool before handling. Heat-resistant gloves or mitts may be an alternative, but these too may be difficult to use.

When Neighborliness Isn't a Virtue

Many college faculties develop positive partnerships with their local secondary schools. It's also becoming more common to accept advanced secondary students into college courses (dual enrollment). These arrangements often lead to requests to provide chemicals or other college materials to a secondary teacher. Resist this means of being helpful. You can't be sure of the training of your secondary peer or of the security system they will maintain for the materials. If an accident occurs, the chain of responsibility may include you.

If you are a secondary teacher moonlighting as a community college instructor, resist the temptation to borrow college reagents and equipment to use with your high school students. If you don't have those materials in your high school storeroom, chances are the facilities are also not properly equipped to safely handle those demos and experiments.

Dangerous Living Through Chemistry

The toxic effect of a chemical is a relative rather than an absolute characteristic. Commonly available household chemicals can be lethal to humans, pets, plants, and the environment depending on the use to which they are put, the concentration of the material, and the means of exposure. Many everyday household chemicals such as rubbing alcohol, bleach, detergents, and other cleaning agents can be toxic and dangerous when mishandled. When you think of the toxicity of a reagent, consider the

WHAT IF . . .

Just because it's chemistry, it's virtually inevitable that certain questions emerge: "What happens when you mix . . ?" and "Will . . . explode?"

Your reply should make clear that it is never acceptable to mix chemicals together randomly. Even minute amounts of some chemicals can result in deadly combinations. Experiments should be the culmination of a careful thought process that predicts results based on extensive preparation and review of data and known characteristics of materials. Steer students to the fascination and excitement of confirming predictions and recognizing discrepant events. As Louis Pasteur said, "Chance favors the prepared mind."

Avoid keeping reagents that can be used to make explosives, especially those that are frequently found in internet "recipes," such as ammonium nitrate. If these items are needed for unavoidable lab exercises, keep them under tight security and store only what you need. If you identify a student with an unusual or inappropriate interest in hazardous chemicals, don't hesitate to make a counseling referral.

6

entire quantity that is available in the open lab rather than the amount a single student lab team needs—the dispensing container could go missing. Don't use, or even order, chemicals in higher concentration than necessary.

The table of selected chemicals that follows shows the dose lethal to 50% (LD_{50}) of subject rodents. These studies are never done on humans, and extrapolation is inexact, but the LD_{50} shows that some common substances—such as PTC (phenylthiocarbamide) paper and nicotine—are very toxic indeed, and some substances that we consider innocuous, such as sodium chloride, can be toxic in high enough quantities.

Exposure by absorption through the mucous membranes may be far greater than by touching. Since chronic exposure is a greater danger for you than for your students, make sure your preparation room is safely ventilated and equipped.

Include some lab activities to give students firsthand evidence of common household chemical hazards they may otherwise ignore. Activities that combine chemistry and biology provide excellent opportunities to discuss real-world consequences of incorrectly using and handling chemicals and drugs.

Poison Control

American Association of Poison Control Center national hotline: 800-222-1222

Sample Toxicities

Chemical	LD_{50}* /kg body mass of rat fed orally
phosphorous, yellow	1.4 mg
Phenylthiocarbamide (PTC)	3 mg
nicotine	50 mg
methyl orange	60 mg
sodium nitrite	85 mg
formaldehyde	100 mg
acetyl salicylic acid	200 mg
cupric sulfate	300 mg
naphthalene	490 mg
lithium chloride	526 mg
calcium chloride	1000 mg
boric acid	2660 mg
sodium chloride	3000 mg

*lethal dose 50 (LD_{50}), the dose that results in the death of 50% of test animals

Note: These data are measures of acute toxicity when all of the substance is fed as a single dose. Toxicity may vary if the exposure is other than by ingestion or over an extended period of time.

The (Not So) Sweet Smell of Success

Remember the old saying, "If it moves, it's biology; if it smells, it's chemistry"? It's probably inevitable that some of your chemistry activities generate unpleasant odors. Odors are a signal that chemicals are in the air. Some chemicals have short- or long-term toxic effects and some can provoke asthma or other severe allergic responses. Be sure you fully understand the nature and toxicity of the chemical that is causing the odor, and do not generate fumes at concentrations that are harmful.

Even when fumes have low toxicity, minimize odors by using small quantities. Plan to conduct labs that generate odor, such as decomposition experiments, on days when the windows can be opened. Teach students the proper way to test for an odor.

In closed or crowded conditions, even slight odors can become major problems, because odors tend to cause excitement and distraction. If the room you are using for lab work was converted from a regular classroom, it may not be as well ventilated as a properly constructed science room. A science lab should have fans or HVAC systems that can exchange the entire volume of room air a minimum of eight times each hour. If your room does not meet this standard, you need to make sure strong odors and fumes are not generated.

Just flipping the switch to activate room ventilation does not guarantee that the system is working properly. As a practical scientist, approach ventilation with a healthy degree of skepticism. Intake vents must remain freely exposed to outside air sources. They often have intake air filters that must be periodically cleaned or changed. In the years since the building was constructed, vegetation may have grown to obstruct the intake. The filter may have been long forgotten or new construction may have limited air flow to the intake. Exhaust ports may become obstructed, limiting air flow. Exhaust fans may not be operating. Often they are belt driven, and the hum you hear may be the electric motor running. But if the belt has broken, the fan blades may not be turning. Check your ventilation system often to be sure it is working properly with sufficient airflow. (Additional information on fume hoods and room ventilation systems is in Chapter 3, pp. 37–39.)

Guidelines for Handling Toxic Chemicals

When considering the toxicity of a chemical, consider the worst-case scenario—that a single individual is exposed to the entire supply available in the lab. Take special note of the toxicity (LD_{50}) of chemical reagents such as dyes and stains that may be meant to be used in minute quantities and pose a very serious hazard if there is accidental exposure to a stock quantity. If at all possible, avoid using highly toxic chemicals—particularly those that pose a risk to the fetuses of pregnant women. If you must, consider implementing these procedures when using a highly toxic chemical:

- Ensure that the MSDS for the chemical is in a prominent and accessible place in the room.
- Label and date every container (including the small ones for distribution to individual groups) with the name of the chemical, its chemical symbol, the word *toxic,* and a description of the hazard in plain English, such as "can cause cancer" or "may cause deformities in fetuses."
- Post a prominent warning sign when the toxic chemical is in use in the room.
- Provide nonlatex gloves for handling and cleanup.
- Protect surfaces from contamination by using trays or pans that can be decontaminated easily.
- Dispose of materials in an appropriate manner.
- Decontaminate the room immediately after use and before another class or instructor uses the room.

6

6

THE SAVVY SCIENCE INSTRUCTOR

Dr. G has been teaching chemistry for 20 years, and every year is as challenging as the first. There was a time when she had only high-performing students. But she has recruited many other students in recent years. She has also become a consultant to the committee developing the new introductory chemistry program, which will include some serious chemistry for nonscience majors. Different students mean different labs and lab procedures.

There are no more bottles labeled *unknown* in Dr. G's lab. Today there are bottles labeled *a* through *d* and four material safety data sheets (MSDSs) available to each student. Students learn quickly that they must treat all four unknowns as if each were the most toxic, corrosive, reactive, or flammable, and that identification information is always available to them in an emergency.

Reagent bottles are smaller these days; so are the test tubes and well plates. Dr. G's students have learned they have to observe carefully and record accurately to find the answers to her laboratory questions. They learn proper disposal methods along with other chemistry lessons.

With all these changes, Dr. G's students have become environmental chemistry experts. They've even developed brochures that instruct on safe handling and proper disposal of household chemicals.

Connections

- American Chemical Society and ACS Board-Council Committee on Chemical Safety. 2001. Chemical safety for instructors and their supervisors. *http://membership.acs.org/c/ccs/pubs/Chemical_Safety_Manual.pdf*
- Cornell University *http://msds.ehs.cornell.edu/msdssrch.asp*
- Fisher Scientific *www.fishersci.com/index.jsp*
- Laboratory Safety Institute, James A. Kaufman, President *www.labsafety.org/about.htm*
- Microscale Gas Chemistry. Ideas for Microscale Chemistry *http://mattson.creighton.edu/LabExperiments.html* *http://mattson.creighton.edu/HowToPrepGases/Index.html*

(cont. on next page)

(cont. from preceding page)

- National Microscale Chemistry Center
  http://microscale.org/
- Samples of Chemistry Accidents. SafeChem.
  www.safechem.com/Campusafe/accident.htm
- Vermont Safety Information Resources, Inc. (SIRI)
  www.hazard.com

6

Striking Gold

Exploring Earth and Space Sciences

The Earth and space sciences may describe the final frontiers, but for many liberal arts students they may be their first and perhaps only trek into these disciplines. Spurred on by tales of adventure, most college students are enthusiastic explorers. But, as astronauts, oceanographers, and geologists know, these frontiers have their dangers. From wearing protective clothing and equipment to using proper techniques with tools, there is much for which explorers must be prepared. Above all, remember there's gold in down-to-earth common sense.

7

I'm Here Because French Was Closed

Modern Earth and space sciences are multidisciplinary, incorporating many of the concepts in life and physical sciences. In some colleges, Earth and space science courses are introductory, single-discipline options. In other places, the Earth and space sciences are woven into courses for nonscience majors and titled Environmental Science or Natural Science. As electives, they attract many students who never considered themselves science-oriented. The range of student ability in these courses is even broader than that in an introduction to a sequence such as chemistry or biology. Whether the course is an introductory, freshman, or capstone course, the skills and habits students learn should help them apply commonsense safe practices to their daily lives.

What's in a Room?

Because of their multidisciplinary nature, Earth and space sciences require most of the facilities and safety features needed in modern physics, chemistry, and biology rooms. (Refer to Chapter 3, p. 33, for more information.) But many institutions lack rooms specifically assigned to Earth and space science courses. So these courses are shifted around to whatever room is available—too often in spaces not properly equipped for safe laboratory work.

If your Earth or space science course must be fit into an existing facility, here are a few features you should try to incorporate:

- *Furniture, such as worktables and storage shelves, strong and stable enough to support the weight of heavy specimens and equipment*—If these are not currently in the room, try speaking with your maintenance supervisor about trading your existing furniture for something more suitable. For example, you may find that older furniture, such as oak tables and workbenches from art rooms or workshops, may work better than modern, lightweight, laminate-top tables. If ordering new furnishings, opt for chemical-resistant resin-topped standing-height tables with bolted legs reinforced by stretcher bars.
- *GFCI-protected electrical service*—Be sure that your electrical supply is ground fault circuit interrupter (GFCI) protected.
- *Hot and cold running water at sinks within the classroom*—If hot water is not available from a central location, get a point-of-use heater installed under a sink. It's essential for proper sanitation in any science lab, but especially one where soils are involved.
- *Safety eye protection for everyone and a means to disinfect safety eyewear between classes*—Many lab activities with rocks and minerals require both impact-resistance and chemical-splash protection. We highly recommend that all safety goggles be suitable for both purposes. (See Chapter 10, pp. 161–163, for a detailed discussion on eyewear and eye protection.)
- *Secure storage space for chemicals and equipment*—You need a secure and separate chemical storage facility (a room with specifically designated cabinets) as close as possible to your point of use. Considerable hazard is created when chemicals from different courses are stored together, especially when the characteristics and special handling requirements of chemicals from one course are unfamiliar to the instructor for another course. We recommend that you order prediluted acids in small quantities, so they can be stored in a small, lockable, corrosion-resistant cabinet. You may also need a fire-resistant flammables cabinet for alcohols and lacquers.
- *Sufficient space for students to work without bumping one another*—Ask students to leave all excess baggage elsewhere if possible. If this cannot be done, establish a storage area away from worktables. Be sure that routes to exits or emergency equipment is neither blocked nor limited by stored baggage. Remove chairs and stools from work areas during labs, and have students work standing up.
- *Ventilation*—Your room should have ventilation directly to the outside that can deliver at least eight exchanges per hour in an occupied lab and four exchanges per hour in an unoccupied lab. A fume hood is required if you use lacquers or other volatile materials. Don't assume, especially when using a makeshift or borrowed room, that the ventilation is adequate; check its operation yourself. (Refer to Chapter 3, pp. 37–39, for more information on ventilation fume hoods.)

Tools of the Trade

Typical Earth science equipment is usually simple, but must be strong and sturdy. As with other science materials, storage must be lockable and secure. Because Earth science equipment tends to be larger and heavier than items for other sciences, storage usually needs to be somewhat roomier and stronger. Because much of the equipment for the Earth sciences is used for cutting or fracturing specimens or obtaining samples under potentially hazardous conditions, you must give students explicit instructions about appropriate as well as safe use. As a general rule, safety eyewear is needed for almost all Earth science activities and should be impact resistant and provide chemical splash protection. The following are some of the tools and equipment specific to Earth and space science lab work.

Eye Protection

We strongly recommend safety goggles that include both chemical-splash protection and impact protection. ANSI Z87.1 is a voluntary performance standard adopted by the Occupational Safety and Health Administration (OSHA) for eye and face protection devices. For example, the lenses should resist penetration from a 44.2 g projectile dropped from 127 cm or a 25.4 mm steel ball dropped from the same distance. Ensure that goggles also have baffles over air holes to protect from chemical splash. Check for the ANSI Z87.1 mark on all eye protection used in the laboratory.

Streak plates and other sharps: Used for the identification of rocks and minerals, streak plates are made of porcelain and can break, resulting in extremely sharp shards. Make sure students understand that streak plates must be handled carefully and must be used on a level surface. Sharp instruments or equipment that may break and produce sharp shards must be handled carefully and disposed of with caution to protect both students and maintenance workers. (See Chapter 10, "Use and Disposal of Sharps," p. 157.)

Rock hammers: Fracturing rocks is not the same as randomly striking or smashing the specimens. Whether in the field or in the classroom, the process always requires eye protection, direct supervision, and careful selection of the specimen. Choose the rock hammer with care—not just any hammer but one specifically designed for the purpose. This is not a place to save money. Poorly made or improper tools can break during use and cause serious injury. Emphasize to students that cleaving and fracturing specimens is a scientifically precise procedure and should be performed only with appropriate caution and training. When possible, we recommend that you purchase specimens that show fracture and cleavage rather than fracturing the specimens yourself.

Rock saws: We do not recommend rock saws for use during regular laboratory activities. They require more attention and preparation than an instructor can give in a whole-class situation. They may be used by an instructor alone or in a small group situation, but students must always be trained and supervised before they use any power tool. You'll need ANSI Z87.1-compliant impact-resistant eye and face protection and gloves, and you will have to make sure your equipment has electric and hand protection. Avoid the use of homemade, modified, and otherwise improvised electrical equipment. Whenever it is available, opt for electrical equipment from reputable manufacturers with UL (Underwriters' Laboratory) certification.

Rock tumblers: Rock tumblers may be used in an Earth science classroom or laboratory for a variety of purposes. However, bear in mind that they must be run continuously for extended periods and may be very noisy. As a guide, refer to OSHA 29CFR1910.95 for federal occupational noise standards at *www.osha.gov.*

Air tools: To do the most gentle work with rocks, minerals, and fossils, geologists often employ a very powerful tool—an air brush. This tool requires special training and eye protection. Use it with care. Automotive repair air power tools have been finding their way into the laboratory. These tools were designed for purposes far different from the classroom and laboratory. We do not recommend using any tool for other than its intended use.

Glues and lacquers: The standard lacquers used for sealing mineral samples have toxic fumes. You will need a fume hood when coating and drying. Use only a very small quantity of the coating material. Chemical-splash eye protection is a must. Avoid using cyanoacrylic glues such as Krazy Glue because of the hazard to skin and eyes. Make sure you have material safety data sheets (MSDSs) on every product you use. (See Chapter 4, p. 56.)

Hydrochloric and other acids: A 10% solution of hydrochloric acid (HCl) is sufficient for the acid test that is standard for identifying carbonate-containing rocks. We recommend that you order prediluted solution in small quantities. The dilution process can be hazardous, and storing concentrated acid represents an unnecessary risk. Store your stock bottle in a locked corrosives cabinet in a secure storeroom. Place small quantities (no more than 10 mL) in fully labeled dropper bottles for use in class. Make sure everyone—doers and observers—wears safety goggles with chemical-splash protection during testing and cleanup. Rinse all specimens thoroughly after testing. Remove jet nozzles from faucets to avoid unnecessary splashing.

Heat sources: Some programs encourage students to dehydrate soil and mineral samples to calculate the water content. Microwave ovens work well for these purposes, but do not use the same appliance to heat food for consumption. If you use hot plates, choose those designed for laboratory use, not those meant for cooking or other household uses. Bunsen burners are seldom needed in Earth and space science. Alcohol lamps should not be used, and we do not recommend them. If you do need this level of heat, check Chapter 3, pp. 42–43, for precautions.

Stream tables and ripple tanks: This equipment provides models of erosion and waves that help students understand what they see in the field. Although they take up a lot of space, they also encourage patient observation. Make sure your stream table or ripple tank is securely mounted so it can't tip. Keep the light sources high and secure. Be sure no one can put one hand in the water and one on the light and that no one can splash water onto the hot lights causing them to explode. Use only GFCI-protected circuits.

Weather equipment: Today's weather equipment is largely electronic, but you might find older instruments, such as thermometers and barometers, containing mercury. These should be eliminated from use and handled as hazardous waste requiring specialized disposal. Many schools have satellite receivers mounted on roofs or towers. Resist the impulse to allow students to climb up to adjust the antennae. Let the trained—and insured—facilities people do the climbing.

Telescopes, binoculars, and optical instruments: These instruments are delicate and expensive and require careful selection and handling. At the college level, students may be able to handle sensitive equipment, but they will still need complete instructions and safety reminders. The most tempting and dangerous error is direct observation of the Sun. See p. 114 for special precautions. Also, be alert to the possibility of transmitting conjunctivitis (pink eye) between users. Anyone exhibiting symptoms of infectious eye irritation should avoid sharing optical equipment. (See Chapter 10, pp. 161–163 for additional information.)

Field and stream equipment: Packing for field trips is both an art and a science. In the field, you will need all of the safety equipment that you need in the laboratory—such as eye protection, MSDS information, secure storage for reagents—and more, including sanitation supplies, weather-related protection, and specialized safety gear such as life jackets. Bring only the quantities you need, pack them securely, and make sure everything is counted before leaving the field site. Choose plastic over glass whenever possible. (Refer to "In the Field," p. 113, in this chapter and Chapter 9 for additional information.)

Rock and Roll

Although the National Science Education Standards (NRC 1996) now place more emphasis on cycles and systems, an important experience for Earth science students is the study of rock and mineral characteristics. The college rock and mineral collection remains a fascinating and inviting source for scientific inquiry. From it, students can learn how to classify and begin to conceptualize the vastness of geologic time.

However, these same collections are usually heavy and can take up a lot of space. Their proper storage and preservation may take a great deal of effort. We suggest that you engage students in some of the organizational responsibility for the collections—especially in

What Part of "No Tasting" Don't You Understand?

Although some geology books or field guides suggest it, rocks should never be tasted for the purpose of identification. Consider the scenario of having students taste a sample of halite to check for a salty taste. Suppose one of the students were an undiagnosed hepatitis carrier, or that the sample were really a heavy metal ore or halite that had been in a box with a chromate. As with all material in a science laboratory, rock and mineral specimens should never be tasted.

preparing specimens for class use and returning them after laboratory work. Keep the heaviest samples on low shelves and make sure boxes and bins are not top-heavy or easily tipped. Although most rock specimens can be stored indefinitely on secure shelving, beware of some specimens that oxidize, such as iron ores. The stains can damage nearby surfaces and materials.

Some ores and elements—such as uranium and cobalt—are toxic, contain asbestos, are combustible at room temperature (white phosphorus) or unstable (sodium and potassium) and should not be used or stored unless very specifically needed—not just for show. If you still have these items in your storeroom, you might first see if a museum or research laboratory will accept them. Otherwise, you need to request that they be removed by a properly licensed disposal firm. Talc may contain asbestos, so, if you wish to keep a mineral specimen, do not allow it to be scraped or crushed. Do not create, use, or store powdered talc, which may contain asbestos. Talcum powder on the market today is actually starch.

The Dirt on Dirt

Although most people associate geology with rocks, the basic scientific material for the course is soil. To study soil appropriately, you need to know the potential hazards in this ubiquitous natural material. Just because it's everywhere doesn't mean it's harmless.

Obtain your soil samples from a source you know, and make sure animal wastes or toxins have not contaminated them. Be sure to research your study site. Industry that was located on the site long ago can be the source for contamination still there. Old smelter sites are not uncommon and pose very serious contamination risks. Old chemical factories or industrial plants may have impregnated the soil with organics and other chemicals that remain at dangerous levels. One of the most common contaminants of outdoor sites is lead paint dust from nearby buildings, bridges, and other structures that have been scraped and repainted.

Soils are also laced with molds, bacteria, and other potential pathogens. Although it would be ideal to use only sterilized soil for lab investigations, it doesn't work for lessons about naturally occurring communities of organisms. Therefore, you must ensure students wash their hands thoroughly after they have examined the organisms in their samples.

Some soil lab activities involve the culture of potentially pathogenic bacteria. Winogradsky columns breed both aerobic and anaerobic bacteria. You should show students the proper washing and cleaning procedures when engaging in such studies, and the contents of the columns should be sterilized before disposal. This activity is appropriate only if you have the facilities and your students have the skill for advanced aseptic technique.

Other protocols involve isolating certain bacteria from soil cultures. In these situations, students may be exposed to pathogens, including *E. coli*, *B. anthracis*, and *C. tetani*. We do not recommend culturing soil bacteria in open culture systems at the introductory college level. "Bioprospecting" for unusual microbes in soil is not appropriate for beginning students. The potential for infection and mold contamination is high. Systems closed with Petrifilm or similar materials may be good substitutes for open culture dishes. See Chapter 5, pp. 74–75, for more information.

Modeling Craters

One often-used lab to model crater formation involves throwing small objects (such as peas) into plaster of paris. Avoid using talc, which may contain asbestos fibers. As with all projectile activities, this requires eye protection.

If you are working with soils in the lab, wash all work surfaces after use. In most cases, soap and water will do. You do not need to use strong disinfectants; the mechanical friction of washing the surface is just as important as the soap or detergent. Do not use a room used for soil studies for preparing or consuming food. In the field, you may want to try the newer hand-cleaning gels made of quick-drying alcohol formulations. However, recognize that these gels are flammable and handle accordingly.

Seeing Stars

Astronomy is an ideal science for teaching the skills that established methods of science—accurate observation, careful record keeping, and proposing new models of the universe. With a few precautions, astronomy can provide a safe and exciting curriculum.

If you are lucky enough to have a telescope or the opportunity to take students to a telescope facility, night sky activities will likely require an evening field trip. Refer to Chapter 9 for information about field trips. But don't think of astronomy as limited to nighttime activities. Many interesting activities that track the movement of heavenly bodies can be performed during the day. The Sun casts shadows that change size, shape, and placement depending on its position relative to the object casting the shadow. Just make sure you take precautions to prevent overexposure to the Sun if you carry on outdoor astronomy activities in daytime. Under certain conditions, the Moon can be seen in daylight.

Don't ever let your students observe the Sun directly. Your first lesson might inform students that Galileo lost his eyesight because he didn't understand the danger.

The lens of the human eye concentrates solar rays in the same way that a hand lens does. Just as a magnifying lens can focus enough heat energy from the Sun to burn a hole in a piece of paper, if you look directly at the Sun, the lens of your eye can focus enough of the Sun's rays to burn a hole in your retina and cause permanent eye damage. Sunglasses do not provide protection for looking at the Sun, and using layers of exposed photographic film is a method too unreliable to be trusted. Neither should you use the inexpensive filter glasses (with cardboard earpieces) offered by some vendors—there are too many possibilities that this type of protection will fail. Teach students the pinhole-reflection method for observing solar eclipses.

Topic: astronomy
Go to: *www.scilinks.org*
Code: SSCC112

In the Field

An Earth and space science course is really incomplete without fieldwork. Whether for a day or a week, on campus or miles away, time in the field should be safe and memorable. As you plan, keep in mind the diverse experience your students have had. Some may be experienced travelers; others may barely know how to navigate across town. Your course should give students a new way of observing the world around them that they may take with them in their travels for years to come.

Even more than activities in the laboratory, fieldwork must be planned meticulously with safety in mind. If your students are novices at fieldwork, you may want to plan some simpler activities on campus before embarking on a more extended or elaborate trip. Go to the site in advance, and look at it as if you were a suspicious stranger. Are there physical perils such as sinkholes, electrical wires, or unstable structures? How about biological threats—insects, snakes, rodents? Will your students be together or move out on their own? If they work independently, what directions will you give them to make sure they can quickly get themselves back to a central meeting area? Careful attention to weather forecasts before and during a trip is also imperative.

Protective gear is usually required for trips. Depending on the site, hard leather shoes with toe protection and ankle support may be needed. In quarries or rocky areas, steel-toed work shoes may be the minimum footwear. You may also need to use protective headgear. Students may be able to borrow hard hats from construction workers if they have sufficient advance notice. Make sure that hard hats aren't merely decorative and that the inner framework for the hard hat is in place.

Some fieldwork requires chemical tests, but this sort of equipment is used less and less. That's because electronic probes are safer, more accurate, and far more convenient. See Chapter 9, "The Great Outdoors," for more information on field trips.

If you plan work in mountainous terrain, be extremely cautious and certain that you and your students are properly trained. Neither staff nor students should ever step onto a glacier or engage in rock climbing without proper technical training. Glaciers are particularly hazardous with hidden crevasses and changing ice quality. Specialized equipment and in-depth knowledge, in addition to the technical skills, are needed before any thought is given to visiting these areas.

WET AND WILD

If you are conducting activities in or near water, water safety and swimming instruction may be needed. Make sure you have water safety equipment, such as personal flotation devices (PFDs, previously called life jackets), in the amounts and types required and that everyone is properly instructed on their correct use. Your area probably has a group of volunteer Coast Guard Auxiliary or Power Squadron members who would advise on water safety.

Mountain streams run fast and cold. They are not a place to take a dip. Extreme caution is required. Large stones make footing difficult in the rushing water. Water depth changes quickly in the transitions between pools and runs. Rivers may not be as swift and cold as mountain streams, but they have their own dangers. Sinkholes and submerged sticks or brush can trap the unwary. Even a weak current can pin a person under water. River edges can have a deceptive bank overgrown with vegetation and deep swift water beneath. Solid-looking ground can give way quickly to a plunge into deep fast-moving water. All work near water requires caution and thoughtful preparation. Each year, stories in the news chronicle events for which enough care had not been taken.

7

Golden Opportunities

When you teach Earth and space sciences, you help students see the world around them with a more discerning and understanding eye. Your Earth and space science lectures and activities present an authentic way to connect theory with the real-world environment. After their course with you, your students may never whiz past a road cut, mountain, desert, or shoreline; or view a quarry or the night sky; or explore a cave in the same way again.

Some travel and vacation adventures may be accompanied by inherent dangers that can be mitigated by lessons learned in your classes and during your fieldwork experiences. When traveling with family and friends, your well-instructed Earth science students should know how to help everyone explore more safely in rugged outdoor environments. Don't teach safety lessons as just a set of rules but rather as a way of life.

THE SAVVY SCIENCE INSTRUCTOR

Ms. D's students have rafted white water, climbed mountain peaks, and dived to incredible depths—much of it courtesy of the United States government. They are carried to these virtual adventures on the wings of satellites using the images and data made available to everyone by agencies such as the National Oceanic and Atmospheric Administration (NOAA), the United States Geological Survey (USGS), and the National Aeronautics and Space Administration (NASA).

Satellite images and other materials also help Ms. D and her classes plan for fieldwork. Before each field experience, her class uses USGS quadrant maps, NASA satellite images, and other data to identify physical features, potential hazards, and challenges of the proposed field site, and to plan "what if" scenarios to mitigate problems well before the virtual trip.

Other data and maps reveal overhead and buried power lines, water courses, and navigation electronics. Some satellite images can even show the moisture content of forest and grassland sites to assist in determining the fire hazard level of prospective field sites.

All this vicarious adventuring helps Ms. D's students get more out of their actual field studies, because, once they arrive at a real field site, they know what to expect and waste little time starting to make observations and record data.

When the field trip is over, Ms. D's students hit the satellite again. They upload their data to the internet to share with fellow student researchers around the world.

7

Connections

- The Microbial World
  http://helios.bto.ed.ac.uk/bto/microbes/winograd.htm
- National Aeronautics and Space Administration
  www.nasa.gov
- National Oceanic and Atmospheric Administration
  www.noaa.gov
- Standard 29CFR1910.95—Occupational Noise Exposure
  OSHA Occupational Safety and Health Administration, U.S. Department of Labor
  www.osha.gov
- U.S. Geological Survey
  www.usgs.gov

Falling for Science

Physics Phenoms

Somewhere among your community college students, there's a tinkerer who dreams of launching a rocket, a mechanic designing a racer in a page margin, a stagehand mixing lights for a play, a sound technician with music reverberating from ear to ear. Each is a physicist at heart. It's our job to show them how the laws of physics play a role in some of their favorite phenomena.

Not Elitist Anymore

There was a time when the only students enrolling in introductory college physics were those already committed to a serious career in science—and who had been tops in their high school science classes. Today, we invite a far broader range of students to stand, as Newton said he did, "on the shoulders of giants," and take physics courses, not just for a major, but as sound preparation for many careers.

Students come to college physics by different paths. In addition to the traditional high school physics course usually taken in the senior year, many secondary schools offer excellent grade 9 and 10 conceptual physics classes with minimal mathematics requirements; others provide interdisciplinary physics programs with lessons on the history and methods of science along with investigative physics. Sometimes vocational education instructors teach applied physics. Older students may bring more practical experience with physics principles than their instructors but may lack formal training in the theory behind their practical knowledge. The knowledge and experience of students in "typical" introductory college physics will be far more diverse than it once was. So you'll have to adjust teaching methods and safety instruction to ensure full and safe participation by all.

Perpetual Motion

Real-world applications of the laws of motion abound and can greatly increase understanding. But, take special care when you plan activities with materials that swing, rotate, speed, or drop from above. Forecast the trajectory if, for example, a tether should fail, a cart run off its track, or a gust of wind catch a falling object. Plan protection, including eye protection, for all people and objects that might end up in harm's way. Ensure that the enthusiasm of a student preparing to drop an object from a high place does not place the student in jeopardy of falling. Never allow experiments that violate regulated speeds, that access restricted facilities such as unprotected roofs, or that involve any other illegal or imprudent actions.

While teaching the laws of motion, you might use the opportunity for students to apply their knowledge to calculations on the motions of automobiles. Stopping distances, inertia, momentum, transfers of energy from one moving object to another or from a moving object to a nonmoving object—all may prove eye-openers to the drivers and pedestrians in your classes and improve safety in students' lives outside your lab.

The Sound of Music

To many college students, sound is what's blasted through headsets and "the wave" is a group activity. As you expand your students' view, consider planning some labs involving noise measurements and take the opportunity to connect decibel levels to hearing and sound-induced hearing loss. Protect yourself and your students, limiting the decibel levels produced during laboratory experiments and generally reducing exposure to loud or constant noises from sound generators and other equipment.

Hearing damage caused by sound that is loud, constant, or both is a growing problem. Among the younger set, use of headsets and boom boxes set at full volume and attendance at concerts with music loud enough to be heard blocks away have resulted in serious permanent hearing loss. In a wide variety of vocations, hearing loss is a serious hazard if proper hearing protection is not used rigorously. Learning to use decibel meters and understanding sound as a powerful source of energy may not only ensure safety in your classroom but also help students understand the necessity for reducing noise and using hearing protection in the workplace. For guidance, refer to OSHA 29CFR1910.95 for federal occupational noise standards. See *www.osha.gov.*

SCILINKS
THE WORLD'S A CLICK AWAY
Topic: noise pollution
Go to: *www.scilinks.org*
Code: SSCC118

8

Let There Be Light

Light shows, tanning beds, and laser pointers may expose students to hazards that they don't recognize. Other students may be engaged in construction, metal work, glassblowing, or other crafts or trades that expose them to ultraviolet (UV) radiation in the workplace. Take the opportunity to provide your students with some classroom applications of their knowledge and help them understand the necessity of eye and skin protection.

As with any laboratory work, ensure that proper protective equipment and devices are in place, that equipment is properly functioning, and that the lowest possible settings and exposures are used. Review "Tools of the Trade" on page 124 of this chapter for more information on specific equipment.

If you work with lenses, consider discussions related to the lens of the eye and proper eye protection. Help students understand the dangers of staring directly at any bright light, particularly the Sun. (For additional explanation, see Chapter 7, p. 111, "Seeing Stars.")

Charge Ahead

Electricity is both a tool and an important content area in physics, but the cytosol of the human cell is a great electrolyte, and electric shock is a constant danger in the home or workplace as well as in the classroom. The severity of a shock doesn't depend just upon the voltage but also upon the current, the body's resistance, and the path the current takes through the body. The median threshold for the smallest current that can be felt is about 5 milliamps (mA) direct current or 1 mA at 60 Hz alternating current (Sarquis 2000). At about 16 mA or more, the muscles of the body freeze and the victim is unable to pull away or let go. At 23 mA, the shock can be fatal. Do not depend on insulating gloves and shoes. The way to resist shock is to avoid it from the outset.

How Much and What Kind of Service?

Classrooms in older buildings are notoriously short of electrical receptacles, and the ones that are there tend to be located in all the wrong places. Don't try to solve the problem by using socket multipliers or extension cords. More wire produces more resistance, less usable power, and a greater fire hazard. If you trip a circuit breaker, you don't have enough capacity. Don't try it again—call maintenance. In the United States, the usual receptacle provides 110 to 120 volts and comes in three types. The oldest type has two equal openings, and plugs can be inserted in any direction (Figure 1). Other receptacles have one slot larger than the other (Figure 2) so that televisions and other electrical devices can only be plugged in one way, not another. Labs should not have either of these two types. The third and preferred form of receptacle has two slots of different sizes and a third opening for a ground pin (Figure 3). The third type is the only form of receptacle that should be in a laboratory.

Figure 1

Figure 2

All outlets should be properly grounded and protected with ground fault circuit interrupters (GFCI). In addition, the room should have a "panic button" or master electric shutoff so that power may be turned off if, for instance, hair or clothing is caught in a piece of equipment. If your room does not have adequate electric service, or if the outlets do not have the proper grounding and safety features, make a request for upgrade or repairs in writing and be sure to include both the

Figure 3

educational and safety reasons for your request. Do not perform unsafe activities or use unsafe equipment until the deficiencies have been corrected. Be sure all electrical equipment you use is properly grounded, and do not circumvent the ground plug with an adapter or "cheater" wire.

SCILINKS. THE WORLD'S A CLICK AWAY
Topic: electricity
Go to: *www.scilinks.org*
Code: SSCC120A

Some devices—such as kilns, stoves, shop equipment, and theater lighting equipment—may require 210 to 220 volts. The receptacles for these lines are generally round, with large openings arranged radially. The arrangement of the prongs in these plugs correlates to the amperage the tool or appliance will draw. Never try to rewire an appliance with a different type of plug than the one that came with it. The amperage to be carried determines the configuration of the socket, so the appliance must be exactly matched to the receptacle intended for it.

SCILINKS. THE WORLD'S A CLICK AWAY
Topic: electrical safety
Go to: *www.scilinks.org*
Code: SSCC120B

Getting the Juice to Where It's Needed

Even if you have plenty of the right kinds of electrical receptacles, you may still be challenged to get the power to the right place. Refer to Chapter 3, p. 39, "Power Up," for information on getting power to the middle of the room. When you plan the routes for your power cords, pay special attention to sinks and water. Never string a cord across a sink or near a water source. Loosely wrap power cords, and place the coil on the equipment. When they are tightly wrapped around the equipment, cords fray at the point of entry and in time will have exposed bare wires and then break.

If you have the opportunity to design a new room, opt for the safest form of receptacle—one that is hidden in a flat, watertight chamber flush with the floor.

Direct Current

Many people consider batteries and direct current a safer alternative to wall current—and they are usually right. But be cautious here, too. Rechargeable batteries are handy but improper charging may damage them, causing a safety hazard. These batteries are most often found in equipment for which they were made. That is their best and safest use, not as a general bench power source. Do not use automobile or industrial batteries. They may leak acid or explode if shorted, and they are usually not appropriate for student or classroom use. If you use a power supply to obtain direct current or to charge computers and probes, make sure it's unplugged after use because it can generate heat and can malfunction.

In Good Repair

Develop a periodic electrical inspection routine for all equipment powered by house current. A dated inventory sheet makes a good checklist and dated record of inspection. Check equipment before use as well as during your comprehensive inspection. Have students also develop the habit of checking equipment before use. Look for cords that are cut or frayed and plugs missing the third, grounding electrode. Inspect more frequently equipment that is fragile or likely to be tampered with.

> **Outdated Rule**
>
> Many older manuals suggest keeping one hand in a pocket to prevent electric current from flowing from one hand, across the chest, and out the other hand when working with appliances. Do not rely on this old trick for protection. Unplug electrical devices before handling them. Stop shocks before they happen.

Repairs should be performed by a competent technician and tested before the equipment is put back in service. When dealing with potentially life-threatening currents, you cannot be doing "good enough" repairs. Wrapping plastic tape around a cut on a cord is not even close to good enough. If a cord is bad anywhere, replace the entire cord assembly and not just the wire or plug. Replacement plugs from the hardware store may be okay for your private workshop but not for a lab or classroom. Should an accident take place and your "good enough" repairs be found to have contributed to that accident, the liability is yours.

Manufacturers often place fuse receptacles on the surface of equipment. Fuse receptacle covers are easily lifted and the fuse removed or tampered with. Inline fuse assemblies are available that may be installed internally—away from tampering.

Have students check electrical receptacles before plugging equipment in. Make sure the receptacles are not partially blocked by some foreign object. Use the test buttons on GFCIs to see that the protection is functioning properly.

The Computer Age

Computers have added a new level of challenge to laboratory design. Facilities designed and built before computers came into use almost never have sufficient electrical service in the locations where needed. In these buildings, surge protectors and universal power sources (UPSs) are needed to protect valuable and sensitive computer equipment from power surges and outages.

Computers and related equipment require additional space. They should not be placed near sinks and should be spaced far enough apart so students can use them without interfering with other equipment and each other. If your room isn't equipped with sufficient power in the right locations for desktop computers, consider laptops that can be recharged at the end of a day and that eliminate the need for power cords and extensions. Wireless computer connections may be an alternative to hard wiring.

The batteries that are used for computers and calculators are usually specialized for the use and the device. If they can be recharged, follow the instructions exactly and use the supplied recharger. Do not mix these batteries with those used for other laboratory work.

8

APPLYING SAFE ELECTRIC PRACTICE

Encourage your students to develop safe work habits around electricity and electrical equipment. Once learned, these habits are useful in the home and workplace as well as in the lab.

- Make sure electrical outlets are not broken or blocked by debris and that plugs fit well in them.
- Never stick anything into an electrical receptacle or an appliance that is still plugged in.
- Use safety covers on unused receptacles where young children might be present.
- Check lightbulbs to make sure the wattage does not exceed the rating of the appliance or wall socket.
- Never use multiple extension cords on the same circuit.
- Do not use extension cords for permanent installations.
- Do not run extension cords under floor covering.
- If an appliance trips a breaker, turn it off until the appliance is checked.
- Keep space heaters at least one meter away from walls and combustible materials.
- Keep combustible materials away from halogen lamps.
- Never use an electrical appliance near water.
- Avoid using electrical equipment during thunder/lightning storms.
- Do not use adapters that bypass the ground or use plugs that have had the ground pin removed.

For more tips: See the National Electrical Safety Foundation website at *www.esfi.org/index.php.*

Radiating Energy

Although you may be justifiably eager to provide hands-on experiences to promote student understanding of radiation and nuclear energy, carefully consider laboratory activities using radioactive materials. Storage and use of radiation sources can pose hazards to the community at large as well as to students and lab personnel.

For introductory classes, we recommend activities that involve investigating radiation already present in the environment. Cosmic background radiation is measurable with most classroom Geiger counters. Students may be surprised to discover radiation sources in smoke detectors (americium) and in basements and well water (radon). Commonly found items such as granite countertops and ordinary rocks may also be low-level sources of radiation suitable for laboratory demonstrations.

Some instructors store low-level alpha, beta, and gamma radiation sources to use in laboratory studies on penetration and distance. If you do so, you must be concerned not only about student use but also about disposal and security. Even if the materials you have are below the levels that require licensure by the Nuclear Regulatory Commission (NRC), store only small quantities of materials in secure facilities. Remember that even low-level radiation sources can have serious consequences if people are repeatedly exposed to them.

Using any level of radiation source in the laboratory requires preparation. Instruct your students thoroughly in advance, and do not admit any visitor or unprepared student during the exercise. Be specific about what students must wear, what they bring to class, and what they do when they are in class. Designated gloves and tongs should be reserved for use only with radioactive materials and stored separately. You must make specific arrangements for disposal and cleanup of exposed materials and have secure storage that no one else in the school community can access for radioactive sources.

The major drawback to storing radiation sources is that they can become an attractive nuisance. Some students may have read about the creation of so-called dirty bombs and become inappropriately fascinated with material you consider to be a routine laboratory supply. When you are planning hands-on investigations with radiation, it's important to consider not only what you want to happen but also the worst-case scenario of what might happen if your radioactive materials were to fall into the wrong hands.

Do not trust older radiation sources found in the stockroom or other places. They could be leftovers from earlier times when nuclear material was easily available. Don't assume. Unless you know for certain what the source is and its qualities, treat it as hazardous. As a general rule, any source that is unknown, inadequately labeled, or obviously a leftover from earlier times should not be used. Isolate and secure it until appropriate assistance can be found to properly remove it.

Students may be fascinated by the true story of painters of watch dials that glow in the dark who were exposed to lethal radiation from licking their brushes (radium). Students can learn from examining X-rays, crystallography, and other products of nuclear science. Many medical and industrial professionals are glad to share what they do with your students and even to guide field trips.

Tools of the Trade

Physical science equipment requires flat surfaces on sturdy tables and counters—usually longer and wider than the standard types for biology or chemistry. Make sure that table legs are fastened to the tabletop with strong bolts and not just glued on. Choose standing-height tables with stretcher reinforcements on the legs. You'll need every bit of 60 square feet of working space per student in the physics room. As in all science rooms, people in the physics venue need eye protection and a means for sterilizing shared eye protection equipment (Chapter 10, pp. 161–163), fire protection (Chapter 3, pp. 43–45), and proper storage and disposal areas (Chapter 4).

In older facilities, storage will take creativity, because cabinets and shelves may not be designed for the larger pieces of physics equipment. It is especially important that equipment powered by house current be connected to ground fault interrupter circuits (GFCI). Even nonpowered equipment may provide a ground path for electrical current from other sources—GFCI protection in all circuits is important.

Here are some other common items:

- *Carpentry tools:* Hammers, saws, and other carpentry tools can be useful for constructing a variety of projects. Make sure everyone is properly instructed in the use of the tools and that everyone, including bystanders, is wearing appropriate eye protection. Choose cutting tools with safety shields, and keep them locked when they are not in use. Instead of duplicating shop tools available elsewhere, consider enlisting the cooperation of a colleague or maintenance supervisor and scheduling classes in a shop or other location where these tools are safely installed.
- *Pendulums:* Safely conducting these activities depends on class discipline and careful attention to the condition of the equipment used. Begin with short strings and rounded bobs. Before each use, check, and have your students check, that the pendulum bobs are securely fastened to their strings. Encourage careful observation and good record keeping. Teach students to begin the swing by letting go of the pendulum bob rather than pushing or throwing the bob. Impact-resistant eye protection may be needed with pendulums and whirling objects such as those used in centripetal acceleration experiments.
- *Momentum carts (cars):* Motion studies may be linear, repetitive, or involve an oscillation of some object. Some motion studies may involve collisions or free-running carts on a variety of track arrangements. All motion studies require careful consideration of placement and potential projectiles. Cars jumping tracks or broken parts from a crash can produce dangerous projectiles. Exercise care to avoid such an accident, and use eye protection.

Make sure ramps do not block walking paths or send cars zooming into the hall. Be sure cars aren't left out on the floor or any other walking area, and put them away when the lab is over.

- *Carbon dioxide pellets:* These can produce significant momentum. Ensure that cars and other equipment are on tracks and not running freely. Use eye protection.
- *Springs:* These items may be compressed, stretched, or accidentally dropped on the floor. A spring under tension is a potential flying missile. They are frequently very sharp at both ends, which enhance the hazard. Even the one on the floor can become a problem when hurriedly stepped on. Use great care and eye protection.
- *Balances:* Keep them secure. They are highly valued for illicit drug activity.
- *Manometers:* Retire any equipment that contains elemental mercury. Treat the mercury as hazardous waste. (Refer to Chapter 6, pp. 61–62.)
- *Siren disks:* The "high speed siren" apparatus often used to demonstrate sound must be firmly installed. Check the safety nut and use only at moderate speed. You should not be generating sound loud enough to make hearing protection for yourself or your students necessary.
- *Tuning forks:* These and other sound-generating devices can shatter glass. Provide clear directions and cautions when using these seemingly simple devices. When studying resonance, put a little tape on the top of a glass tube to reduce the risk of shattering. Use impact-resistant safety eyewear.
- *Ripple tanks:* Set these up in a secure place and away from unprotected electrical outlets. Position lights so they can't be splashed with water. Use GFCI-protected circuits.
- *Oscilloscopes and other large instruments with screens:* Older models of these instruments are often bulky and heavy. They are often placed on shelves above the workbench. Be sure that the instrument is firmly and securely placed on its shelf and that the controls are easily seen and accessed. The height of the shelf should place the instrument close to eye level so that it may be easily and accurately manipulated and read.
- *Cathode-ray tubes (CRTs), including old televisions and computer monitors:* Do not open televisions or computer monitors. They have capacitors that may store large charges even when unplugged. Picture tubes and other cathode-ray tubes (CRTs) may also retain charges and implode.
- *Photogate timers:* A commonly used measuring device, its three modes of measuring (gate, pulse, and pendulum) and calibration must be set prior to use. Its small size makes it vulnerable to being dropped, knocked over, and damaged. Position this and other small instruments securely and not in a tangle of probe and power wires on the bench.
- *Signal generators and other meters:* Signal generators and various multimeters—digital and analog—fall into a similar category. Calibrate meters to measure various electronic values such as voltage, capacitance, current, and resistance so that the appropriate range of measurement is selected. Handle these instruments carefully. Do not pile these small devices amid tangles of probe wires.

- *Variable power supplies, voltage transformers, and generators:* Variable power may be supplied by freestanding equipment or built into the workbench. Ensure that these variable instruments are correctly set up for your particular use. Power supply equipment that is plugged into house current has the potential to supply very large quantities of energy. If you use a voltage transformer or generator for experiments, make sure it does not generate current that is dangerously high. When a small voltage is required for a lab activity, check to ensure that a carelessly adjusted supply does not deliver a substantially larger jolt.
- *Van de Graaff generators, Leyden jars, large capacitors:* Consider eliminating these items. The high voltages generated or stored can produce serious shock. Make sure that no one in the area of a demonstration with these devices has a pacemaker or other medical condition that would make the output potentially dangerous.
- *Lasers:* Laser pointers, radar guns, and the use of lasers in scanners have made the technology seem so commonplace, we may forget the hazards lasers can pose and the precautions that must be taken. Never consider any laser beam or laser-associated instrument to be harmless. Always take precautions to protect eyes and other exposed tissue. Provide structured laboratory investigations with specific directions for data collection and appropriate eye protection throughout. Make sure the target for your laser is set up well in advance, is nonreflective, and is secure. In introductory college classes, do not exceed 0.5 milliwatts in a helium-neon laser.

  The coherent beam of laser light must be controlled. That control begins with the laser itself. Be sure to tape down power cords supplying the laser. Fix lasers in the desired position so that they are not easily moved. Limit the effect of a passing student's bump or stepping on support cables. Be sure that everyone knows the planned beam path and ensure that the path is indeed the beam path. Require everyone in the room to wear laser glasses. For guidance, see laser standards and safety information at *www.osha.gov/SLTC/laserhazards/index.html.*

8

- *Black lights:* These and other ultraviolet (UV) radiating devices should be handled with care and appropriate eye protection. Discuss UV radiation, its many sources, and its potential to cause skin cancers, cataracts, macular degeneration, and other eye damage.
- *Strobes:* If planning to use strobes, recognize that some uses can trigger seizures. Provide adequate information and warning to students or bystanders who may have medical conditions that would make strobes a hazard.
- *Magnets:* With the increasing availability of very small, but very strong magnets, you must be careful that these harmless-looking small objects do not wreak havoc on sensitive instruments, data disks, and other electromagnetically vulnerable objects. Separate and mark or label them in such a way that everyone is aware that they are magnetic and not just a benign object. Be careful that you store magnets so that

they do not disrupt nearby materials or negatively affect each other. Some large industrial magnets were once coated with asbestos. They should not be used and should be regarded as hazardous waste. This is another time to remind yourself never to accept donations. "Free" can be very expensive.

- *Ball bearings and other small objects:* Tripping hazards from such items should be obvious but can be avoided only if you take care when using and storing these materials. Build in adequate time to distribute and collect equipment for all lab activities. To prevent loss and accidents due to misplaced objects, get a complete count of all materials at the end of an activity.

This list is far from complete. The truth is that physics teachers love to use old tools and invent new ones. It's essential that each tool be used with caution. It's the invented demonstrations and tools that are most likely to cause accidents.

THE SAVVY SCIENCE INSTRUCTOR

Ms. T's physics students are also performers. The highlight of their work each year is a physics extravaganza that becomes a traveling show and student-recruiting tool when presented at neighboring secondary schools.

Each extravaganza begins with a light show. But this one involves audience participation. As each colored light beam is added to the display, the audience members try to predict the results and learn about light mixing.

The heart of the show involves physics "magic" tricks. Students pull tablecloths out from under fine china, turn virtual objects upside down with lenses and mirrors, and change sounds by waving their hands. For each trick performed on stage, the "magicians" provide the audience with an explanation of the physics behind the fun.

Inventors display their self-built musical instruments and accompany the show with familiar songs on very unfamiliar devices.

After each show, Ms. T's students join the high school kids on the playing field and share their enthusiasm for physics.

Connections

- American Association of Physics Instructors. Safety in Physics Education. 2002. *www.aapt.org/Store/products.cfm?ShowAll=1*
- Ground Fault Interrupter Circuit (GFCI) *www.electrical-contractor.net/ESF/GFCI_Fact_Sheet.htm*
- The International Electrical Safety Foundation *www.esfi.org/index.php.*
- Physics Safety Issues. The Catalyst. *www.thecatalyst.org/hwrp/safetymanual/index.html*
- U.S. Department of Labor, OSHA—Laser Hazards, standards *www.osha.gov/SLTC/laserhazards/index.html*

The Great Outdoors

Field Studies Near and Far

Field studies and trips to off-campus locations are frequently critical components of science course work. Even with extensive technology and internet support, certain things cannot be accomplished in the lecture hall or standard laboratory. Whether it is visiting a working lab in industry, examining a unique exhibit or collection at a museum or nature center, or observing and collecting data in the field or at a research station, there are occasions when just viewing a video won't do. With any outing—for the entire class or just a few of the students, on-campus during a standard laboratory period or at a remote location for a week or more—the difference between success and failure and between a safe, memorable learning experience and disaster is careful planning and preparation.

So That All May Gain

Because of the diversity of students enrolled at the community college level, you need to consider the physical, financial, and time requirements of your proposed excursion to ensure that everyone in the class can participate fully and safely. Make no assumptions about physical abilities or maturity. Ask and, if necessary, plan preliminary activities that will allow you to observe and assess just how much practice with tools, equipment, and techniques and how much training and physical conditioning must precede a trip. Determine also the maturity and personal responsibility you can count on from each student. This should inform the level of preliminary work and supervision you must plan.

If there is disparity in the physical abilities and maturity of the students in your class, you may need to plan "trips within trips," so that the demands on each individual or team are appropriate to their abilities. You may need to arrange reduced or alternative investigations for those who cannot manage all the activities. In some cases, you may need to provide a trained assistant or modified equipment for students with identified disabilities.

Students who are significantly challenged physically, mentally, or emotionally require thoughtful consideration and decisions. This book cannot possibly present all the dos and don'ts for field experiences and is not meant to do so. But careful consideration, in-depth research and understanding of what you are planning, and a healthy dose of reality are needed to carry out a successful field experience. Even then, the unexpected can happen.

Provide students with preliminary information on your plans and allow plenty of opportunity for individual students to approach you privately to discuss conditions that may limit their participation. If you would like students to participate in an experience that may be physically or financially out of reach for some, make it a voluntary experience during a vacation period or following the end of the term and make sure that students who cannot participate are not penalized.

A Journey of a Thousand Miles ...

Despite the fact that your college students seem like savvy travelers, it's a good idea to choose a venue right on campus or close by for your first fieldwork. Plan an activity that can be completed within a single lab period and that permits you to observe each student carefully so you can gather information to help you plan more complex fieldwork. Structure your fieldwork so that each subsequent experience gets longer, farther away, or involves more complex tasks.

Before setting out to a distant field location, have students practice skills and techniques in local settings. Plan training activities before the big excursion. Consider using simulations—many websites offer great practice—to let your students hone their technical skills so that their work in the field garners reliable data. Here are some local field studies you might consider:

- Use quadrat analysis to survey biota in microenvironments around the campus.
- Gather materials for soil analysis from lawns and plant beds around the buildings or on the college field. This should also be an opportunity for students to practice gathering materials without damaging the site.
- Use school records and photographs such as yearbook archives to compare the relative effects of weather on metal bleacher components, bicycle stands, and building exteriors. Have students begin the documentation of changes in the campus landscape, buildings, and vegetation, and use the collected data from year to year.
- Find the number of moles of sidewalk chalk required to write the name of the college in 10 cm high letters.
- Observe and estimate the momentum and inertia of children on the neighboring elementary playground equipment.
- Determine the energy flow and heat loss through a section of outside wall at differing temperature gradients. Apply this information to heating and cooling requirements, and determine efficiency.

Whether your first trip is just across campus or to a city park down the street, take time to observe the physical abilities of individuals to keep up with the group and the ability of students to work responsibly and independently. Use this information in planning further activities.

Preview and Prepare

You must thoroughly examine any field site you are considering before your trip. You need to take with you tentative plans and a thorough checklist to find the potential hazards that aren't obvious from looking at the venue's website or reading program literature. After you've examined the site, write your final plans, including safety preparations, assistant training, clothing and gear requirements, and schedules for travel, work, and site cleanup.

Who, What, When, Where

Depending on the topography of the site, you must plan a method for monitoring where each student is, what students are doing, and calling all students together at designated times and in emergencies. The class is not ready for a trip until everyone knows what action is expected even when the unexpected occurs. Put particular emphasis on the rule that no one may leave the site and his or her assigned location without explicitly informing you. Immediately report anyone missing, and stop all other work until the missing person is located.

Expert Help

If you expect to use staff or guest experts from the field site, make sure you have a full and detailed planning session that includes a discussion of exactly how you will prepare your students, what you expect your partners to do with your students, and enough information about your group and individual students for an outsider to know what to expect and how to work safely and effectively with your class. Don't assume that just because your guides are experts in technology, wild plants, or rock outcrops they understand the needs of students. A good planning conference is worth its time in gold.

Permission

Make sure you have written permission to be at your site and to do what you plan to do. If you are visiting public lands and working with plants or animals, be sure you have all the permission and permits necessary. Some plants and animals are listed as endangered and may not be disturbed. You and your institution could be considered liable if your students pick a plant or filch a fossil without authorization.

9

Medical Records

You will need medical information, especially on extended trips and trips to remote locations. Explicitly ask each student to write down for you the medical information necessary to ensure his or her safety on the trip. This includes information concerning medication and conditions that might limit their participation in the trip. Give students activity outlines with sufficient detail that they may consult with their personal physicians regarding participation. If there are minors in your group, then the medical information

> **Waivers of Liability**
>
> Your institution may require that students sign waivers of liability before participating in field studies and trips off campus. Check with the administration about this, but also beware that waivers of liability can be overturned easily in a court case in which your negligence can be established.

and specific permission to participate must be obtained from a parent or guardian. For minors, you must be sure also to have emergency treatment permission slips.

You must carefully guard any medical information provided to you to ensure that no student's privacy is compromised. Keep all written information in a secure place where only those with a need to know have access during the trip, and securely destroy the documents following the successful completion of the trip. See Chapter 10, p. 169, for more information regarding confidentiality.

Structure and Schedules

Despite their apparent sophistication, college students need structure on a field trip. The type and degree of structure will depend both on the type of student and type of field trip. With a class of responsible students in a relatively benign indoor environment such as an aquarium, individuals or groups can work on assignments independently and return to a prearranged rendezvous. But a trip to a geologic site or an overnight trip calls for considerably more safety checks and a closely monitored buddy system to keep the participants and the integrity of the program safe. Each trip should have a written schedule that includes a precise time of departure from the campus or time of meeting at the site. The schedule should include the time the group is expected to depart the site or return to campus. Prior to departure, establish a telephone chain for informing a contact person on campus and others, including parents and guardians of minors, of changes in plans or a delay in the return.

Buddies

Establish groups, with specific group responsibilities, in advance. Though students may be assigned to work in small groups, for added safety, you should pair students and ask that each member of the pair be responsible for keeping his or her buddy in sight at all times. In the excitement of activities, an individual student might not be missed, even in a small group. This buddy system is especially important for any field studies at or near bodies of water.

Students as Stakeholders and Representatives

Although supervision is important and required, you have no guarantee that each participant will act in an appropriate manner at all times. The more you can engender a sense of ownership of the program among your students, the better off you will be. If all the students in the group are prepared to defend the integrity of the activity from any individual participant's reckless behavior, the less likely such behavior will go unchecked.

Well before announcing a field trip, make sure your plans conform to college policies, and be sure that the administration will support your rules with firm consequences for any participant who breaks them. Establish the perimeter of travel, the expectations of behavior, curfews, and consequences for substance abuse. Then prepare students carefully. Provide written and oral precautionary instructions, and document all of your preparation. If you have students who may present a danger to themselves or others, then you must decide whether a proposed field trip should take place at all.

Every student (and parent in cases of students under 18) must be thoroughly informed of expectations and agree to abide by rules of conduct for the trip. Consequences for infractions must also be agreed upon ahead of time. Individual students as well as the group as a whole are representing your institution as well as themselves. All students should sign commitments to attend, with rules and schedules, in advance. For students younger than 18, a formal permission slip must be signed by a parent or guardian.

HARD DECISIONS

Not all students can exercise needed self-control at all times. Some make purposeful decisions to perform as they wish, and others inadvertently fall out of line. Student behavior and casual remarks can be very damaging when the students are off campus because outsiders may generalize an incident or remark to draw conclusions about the institution you come from. If a student purposefully becomes a liability to the trip and program or an unthinking act jeopardizes the group, you as the leader must make difficult decisions.

If a student has a record of being disruptive, you must make arrangements for appropriate monitoring. In some cases you might need to exclude a student from the trip. This is a difficult and possibly legally challenging decision. The requirements for appropriate behavior and the possible consequences should be in your syllabus. Keep a lengthy record of past disruptions and your response to them. It is important to have a carefully dated and documented record of your previous interventions and attempts to work with any student requiring special attention. If the student is underage, further considerations are required. Work closely with your administration and student support staff before the event.

In some cases, you can avoid long arguments and appeals by offering equivalent written work for those who should not be allowed in the field. Unfortunately, there may be times when a field trip is best left for another semester or another year.

Permission Slips in a College Class?

Dual enrollment and younger students may be minors. If you have a student younger than 18, you must have a permission slip signed by a parent or guardian for the student to take part in a field trip. Be sure to have emergency treatment permission slips for any minors in your group. Should an accident occur, this simple precaution will be invaluable to you and the injured party.

Expected time of return should be clearly stated, and parents and guardians should be asked to indicate what arrangement has been made for the student upon return. Just like the clerk at the local package store, do not rely on students' looks or assurances.

Overnights

For overnight and long-distance trips, there are further considerations. To minimize disruption to your own and your colleagues' academic programs, try to schedule overnight trips during weekends or school vacations. Make sure that participating won't be an academic or financial hardship to students. You might think first about the opportunities near home rather than the faraway places you would like to see. Avoid scheduling an expensive trip that is free for you because you are the organizer.

Some of your students may be in our country on visas and have travel restrictions. Some may still be minors and require parental permission or as much as 24-hour-a-day monitoring. Some may have handicaps they have not documented but which might become relevant on a long trip. You can establish an e-mail box at which students can tell you privately if there are any issues that might keep them from fully participating in a trip. Make sure that you have all of the medical information you might need, including information about the medical facilities closest to the site.

All students, regardless of age, should be explicitly prohibited from leaving the group and venturing forth alone, especially after hours. Minors must not be allowed to be apart from the group unsupervised under any circumstances.

Another consideration is what to do with a participant who violates the codes of conduct that you have established ahead of time. If possible, such infraction should result in immediate exclusion from the group and return home at the participant's own expense.

Help With Extended Travels

Before long-distance or overnight travel, get the assistance of an experienced travel planner. This may be an institutional requirement. Some institutions may have in-house travel planners or a travel-planning course, which can provide assistance.

Foreign Travel

For foreign travel, check U.S. State Department travel advisories at *www.state.gov/travel*.

Consult with hospital travel clinics and check Centers for Disease Control and Prevention (CDC) bulletins for immunization information. *www.cdc.gov.*

For information on backcountry work see "Into the Backcountry With Care" in this chapter on p. 141.

Help?

As with regular lab investigations, you are responsible for everything that occurs during a field study. This includes training and directing all assistants and consultants.

If you have lab assistants for your trip, it is your responsibility to prepare them thoroughly for their duties. If lab assistants are not available, you might solicit the voluntary help of former students who have successfully completed the same field study. Bring sufficient trained help to avoid exhausting yourself. If you plan to use an outside expert, do not assume the expert knows how to work with your group. You will know best how to interact with your students.

Regardless of whom you bring along for assistance, make sure you have reviewed the entire plan, purpose, and procedure for the activity and assured yourself that all your helpers are fully qualified to provide responsible assistance. Guest instructors must also be well informed and understand your teaching style and goals. If you want students to observe and draw their own conclusions from personal discovery, the last thing you want is an assistant or guest expert who spews forth everything he or she knows about anything observed.

Extra untrained participants, such as friends or children of students, should be discouraged. It is enough responsibility to plan for students whom you have observed and instructed on a regular basis. Adding individuals whose skills and character are unknown to you is an invitation to trouble, especially if it adds a social dimension that distracts from attention to the planned investigations.

Who's in Charge?

Regardless of how many assistants and experts you use during your field trips, the ultimate responsibility and liability rests with you. It is important to set up procedures so that students take on substantial responsibility for the quality of their work and conducting themselves professionally, but doing so does not relieve you of the duty to check and intervene.

How Are You Going to Get There?

The policies of institutions vary, but because most community college students can arrange their own transportation, it is typical to schedule local field trips in a way that students get there on their own. It is important, however, to recognize that community college classes may include students too young to drive, students who carpool to classes, and students who are dependent on public transportation. If your class is going to meet at a library, a park, or a museum instead of on campus, make sure everyone has a way to get there and a way to return. Provide a map and a phone number for emergencies, and let students know how to tell you privately if there is a problem.

There may be serious liability issues for the college if students drive their own or other vehicles to a school-sponsored field trip. Consult with your college administration and legal counsel as well as your own attorney for information, instructions, and required paperwork. Don't drive students unless you are licensed and insured specifically for that role.

Public transportation—buses, trains, trams, trolleys, subways, and ferries—may be an economical option for individual or group travel. But don't assume that all of your students know how to navigate a public transportation system. If you are traveling as a group, remember that keeping a large group together for boarding and disembarking is significantly more complex than with individuals or small groups. Check explicitly to make sure the transportation system is prepared to accommodate your group. The regularly scheduled run may not have enough room for more than a few extra passengers. Make sure you review embarking and disembarking instructions with students and assistants. Provide explicit instructions on what to do if someone is left on the platform or in the vehicle.

If you are planning to walk or hike, make sure all participants are conditioned and that they have the correct footwear. Whether traveling in small or large groups, the slowest walker or hiker should be at the front and a designated person at the rear should check in regularly to ensure the group stays together. The pace of the entire group should be set at the speed of the slowest participant. No member of the group should be permitted to pass the designated lead or fall behind the designated rear walker or hiker.

Before departing from any field or off-campus site, conduct a check-in so you can account for everyone at the end of the activities.

HOW MANY IS ENOUGH?

There is no clear ratio of instructors and assistants to students that can be applied to fieldwork. The right number depends on such disparate factors as the distance and location of the site, the hazards at the site, the nature of the activities you have planned, and the experience and maturity of students in your class. Here, nevertheless, are some guidelines:

(cont. on next page)

(cont. from preceding page)

- Do not count yourself in the instructor-student ratio or assign yourself to a specific group. You need to be available to monitor the overall activity and support the assistants and outside experts you have along.
- Do not count aides or disability assistants in the instructor-student ratio. In groups with special-needs or disabled students, their assistants should be considered additional help rather than a primary resource.
- If a student has a record of being disruptive, you must make arrangements for monitoring. In some cases, you might need to exclude a student from the trip. Work closely with your administration to be certain you are not violating a special-needs or Americans with Disabilities Act (ADA) regulation.
- Most trips require at least one trained assistant. If there were an accident or a lost student, you would need someone to stay with the large group and someone to handle the emergency.

Regardless of the number of instructors and assistants available, students should understand they are responsible for conducting themselves safely and professionally.

Museums, Zoos, and More

A visit to an informal science center is most effective when advance preparation is thorough. The most productive and the safest of such visits have a narrowly focused purpose that has been discussed in advance with the educational staff of the institution to be visited. The facility may have preplanned exercises you can modify. If not, visit first and develop your own.

Preparatory classroom work before the visit is important. The greatest potential for disappointment and trouble arises when the visits are general tours or when the instructor turns students over to the institution staff. You cannot turn over your responsibility for the instruction and motivation of your students to some other person. If students do not have a specific series of tasks to complete and objectives that are an integral part of your program, they can be easily tempted to race through the site, cause disturbances, or simply waste the time. With a clear focus, students are much less likely to amuse themselves in unproductive or dangerous ways.

Scout It Out

- What are the natural boundaries of the site?
- Are there hazards within the boundaries of the site?
- Are there hazards accessible from the site?
- Is there any toxic vegetation?
- What is the likelihood of insect hazards, such as deer ticks, mosquitoes, bees, and spiders?
- Are animal encounters likely? (Coyotes, foxes, rodents, and even moose are becoming increasingly common even in urban locations.)
- What are the tripping or falling hazards?
- Can you see and monitor all students at all times?
- Are there private property or other trespass issues?
- Is there a chance of the presence or intrusion of other groups or individuals?
- What are the conservation restrictions?
- How much sun are students likely to get?
- Are there water hazards?
- Are there restroom and washing facilities? If so, where?
- What is the nearest emergency medical facility?
- What is the nearest source of help?
- Where is the nearest telephone? Is there reliable cell-phone coverage?

9

Outdoor Sites

Whether the site is on campus, just outside the classroom doors, or far enough away to require a bus or an overnight stay, you need to check out the possible hidden hazards, especially if you are using an unfamiliar location. You should also make sure the site does not carry restrictions for use and access. Ask for written permission to do your studies on the site. Check this particularly with conservation land, wildlife preserves, and private property, and make sure there is no hazardous materials contamination.

We Have Met the Enemy and They Are Us

If there is a structure such as a bridge or tower near your chosen site, anticipate hazards. Do not count on students, especially those focused on the field study, to read and obey "Danger, Keep Out" signs. Identify the hazards, and clarify the limits of investigation. You should find out if refurbishing projects have taken lead paint off a structure and allowed lead dust or asbestos to contaminate an area. If utilities have rights-of-way in or near a site, identify any high-voltage hazards. Sites near utilities and manufacturing and research facilities should be checked for toxic wastes. Areas formerly used for military training may contain unspent munitions. Turn these possible hazards into a safety learning experience, and do not use any site unless you can be certain that you can keep students a safe distance from hazards.

Water, Water, Everywhere

Some of the most common field studies involve water sampling. Check your site in advance. What is the footing like near the water's edge? Is the water biologically or chemically contaminated? Will there be mosquitoes or other biting insects? Are there snakes, alligators, snapping turtles, or other potentially harmful organisms?

Is the water deep enough that you'll need water safety equipment and clothing such as waders and gloves? Can all participants swim? If so, how well? Drowning can occur in very shallow water, especially if someone is distracted or loaded down with equipment. Regulation personal flotation devices (PFDs) should be worn rather than just available. In the worst case, who would be responsible for water rescue or cardiopulmonary resuscitation (CPR)? For trips that require a boat, make sure the operators of each vessel meet all licensing, safety, and insurance standards and requirements. Students should not be permitted to operate boats. Consult a flotilla of the local Power Squadron or Coast Guard Auxiliary for help in getting trained and reliable boating support.

Water studies may use chemicals or probes. All equipment used on a field trip should be tested in the classroom first. Practice the procedures in advance. Incorporate responsibility for maintaining and operating the equipment in group tasks and procedures. If you use chemicals, you will need to bring material safety data sheets (MSDSs) along. Establish rules for where samples may be collected and the depth of the water in which students can explore for them.

For work in areas where tides change water depths rapidly or significantly and areas where rip currents are a possibility, you must obtain precise up-to-date information on the timing and nature of tidal changes and currents. Students must wear proper PFDs and be tied to an anchored point on shore if possible. Do not tolerate any infractions.

An Animal's Home Is Its Castle

In the field, your students will trespass into the habitats of animals. Teach them to respect these homes and the ecosystems they visit. Prepare them by asking questions and perhaps assigning reading and research in ethology: What animals are likely to be found at or near your site? What is the normal behavior of these animals? What are signs that the animal may be sick or injured? You do not have to be the expert, but you do need to check with a naturalist or guide who is familiar with the location and who can advise you thoroughly and accurately.

As a rule, students should not approach any animal—living or dead. The normal behavior of animals is to hide or run from humans. One that approaches your group or does not scurry away is more likely to be sick or injured and should be left alone.

Marine Microcosms

Rocks on marine beaches provide many habitats. Organisms making their homes on the top are different from those found on the sides or bottom. After lifting a rock to inspect its population, return it in exactly the same way it was found—top up and bottom down. Students carelessly lifting and dropping rocks can destroy a beach's living systems.

Snorkeling is a great way to observe marine ecosystems. But it requires thorough preparation, lifeguard support, and special cautions. Coral can be extremely sharp and dangerous, even when students are in knee-deep water. Again, preparation is the key.

Teach students to avoid nests or dens: they are likely to be the home of fleas or other fauna best left where they are. Rodents in some areas harbor fleas carrying plague bacteria, *Yersinia pestis*. An animal protecting its young is likely to be very aggressive. Above all, do not touch or approach a sick or injured animal—do not attempt a rescue or try to bring it back to your classroom. This is usually illegal and often dangerous. Dead animals may carry organisms that can transmit zoonotic or potentially zoonotic diseases, such as Lyme disease, encephalitis, rabies, and anthrax, and H5NI avian ("bird") flu. (See Chapter 5, pp. 78–79.)

Rocks and logs provide shelter for many organisms. A single rock or log may be home to dozens of plant and animal species that may live in, on, or under the object. Students should be instructed not to disturb these habitats or, if examining them, to exercise care to limit movement and exposure and to be certain to return the object exactly to its original position. A class traversing a beach or forest carelessly kicking, rolling, and lifting rocks or logs can be devastating to native flora and fauna.

You need to be aware of insects that are indigenous to the area and whether they could carry infectious human disease, such as Lyme disease, Rocky Mountain spotted fever, West Nile virus, eastern equine encephalitis, hantavirus, dengue fever, and malaria. Check with local medical services if you have any questions or doubts. To ensure that allergies and personal preferences are treated appropriately, ask students to supply their own insect repellents. Students need to understand the repellents they use may affect others, and so they must apply them sparingly and away from others. Determine whether anyone on the trip has allergies to bee stings or other insect bites and what you are required to do about it. See Chapter 10, "First Aid," pp. 158–159, for information about using EpiPens to forestall an anaphylactic reaction.

Parsley, Sage, Rosemary, and Thyme

Vegetation in an outdoor area can also pose hazards. Remind students in writing that nothing should be tasted or eaten.

SCILINKS
THE WORLD'S A CLICK AWAY
Topic: allergies
Go to: *www.scilinks.org*
Code: SSCC140

Pollen and spores may cause allergic responses. Be sure to check for allergic sensitivities among your students and assistants. If you have sensitized students, you may want to avoid outdoor activities altogether when pollen and mold counts are high.

Plants can cause serious irritation on contact. The best known are members of the *Rhus* family, commonly called poison ivy, poison oak, poison sumac, and poison elder. These plants are widespread in outdoor areas and may have different appearances in different habitats and seasons. Learn how to identify them and teach your assistants and students to do the same. If you expect to encounter *Rhus* species on your trip, carry products that can be applied immediately if skin contact is made with the plants.

Some people mistakenly believe they are immune to the irritants in these plants because they have come into contact with poison ivy or its relatives without developing the classic itching and blistering response. But sensitivity to the antigens can develop as a result of a series of exposures, with each subsequent contact resulting in a stronger response.

The saps of many plants are serious irritants, particularly milky-looking saps. Students should be taught to avoid touching plants they are unfamiliar with and to wash off thoroughly following accidental contact. Be sure to warn against rubbing eyes, which can transfer substances from the hands to the eyes.

Campfires are sometimes included in overnight trips. However, it is best practice to avoid them. The safety issues include the potential of accidentally starting an uncontrolled fire and the production of highly toxic fumes if the wrong wood or twigs, such as oleander or *Rhus*, are included. Additionally, the forest needs the branches and twigs that you burn in a fire to recycle nutrients through the ecosystem. Wood left as it falls re-enters and recycles. Wood burned in a fire results in its nutrients, especially nitrogen, being lost in the smoke.

Plants are a vital part of an ecosystem. Just as animals are to be treated with care and respect, so too should the plant community. Leave no trace of your passing.

SPF

The potency of sunscreens and sunblocks is measured by their sun protective factor (SPF). Sunscreens with SPFs between 15 and 30 will block most UV radiation, according to the National Cancer Institute.

The Sun Also Rises

Tanned skin was once associated with health, vigor, and glamour. But we now have serious cause for concern about skin damage caused by ultraviolet (UV) radiation during sun exposure. Excessive UV exposure when young greatly increases the risk of skin cancers years later, and recent data show an alarming emergence of melanoma in teenagers. Preparation for outdoor fieldwork can be the perfect opportunity to discuss current knowledge about the harmful effects of exposure to UV radiation from natural and artificial sources.

Hats, long-sleeved clothing, and sunblock are necessary precautions for everyone working in the outdoors. Be sure lips and ears are also protected. Cumulative UV exposure is also thought to be harmful to eyes, so UV-blocking sunglasses may also be in order. If students will be in the sun for several hours, consider SPF-rated clothing.

Also consider heat and dehydration. Make sure the work area does not get too hot—or too cold—and that everyone drinks plenty of fluids.

Into the Backcountry With Care

Extended trips or overnights, including packing into remote areas, can be rewarding and one of those learning experiences that stays with students, positively, for the rest of their lives. Memorable experiences come from trips that are well thought out and well planned, trips in which students have participated in the entire process.

SURE FOOTING

Sturdy protective footwear is a must at most field sites and for trips that involve substantial walking or hiking. Arch and ankle support are essential. Many styles of sneakers do not provide sufficient protection. Steel-toed shoes are often required for rugged terrain. Flip-flops and sandals are not appropriate for beach work; special beach shoes are available at very low cost.

Warn students and assistants that new shoes must be broken in before use on an extended walk or hike.

Where you step is also important. Twisted ankles are frequently the consequence of slipping on what appears to be a solid surface. Warn all participants to be especially cautious of stepping on rocks or logs with dark-colored surfaces that may be slippery from moss or algae growth. Be aware that stones and boulders can work loose and fallen branches and logs can roll. When people are standing in or wading through water, discerning whether the footing beneath is uneven or soft is difficult, so everyone should test the footing with careful and slow steps.

Leaders

An extended hike or backpacking trip requires at least two leaders—you and at least one other instructor or senior assistant—who work closely together and are experienced. One leader should be at the front and the other at the rear. The leader at the front sets the pace dictated by the slowest member of the group. Make frequent rest stops, so that the instructors can assess each individual, exchange information, and make needed modifications. No one should ever be allowed to fall behind the instructor at the rear of the group. In an emergency, the rear instructor must be able to use a prearranged signal to stop the entire group.

Expectations

Should your field trip be into a more remote area and require packing in supplies and equipment, there are further considerations. Every person taking part must thoroughly understand what will take place, the physical exertion required, and how remote the venue is—perhaps a full day's hike from outside assistance and lacking cell phone and radio communications. Take special precautions for the physical ability and reliability of student participants. Review "Hard Decisions" on p. 133 if you are considering backcountry excursions.

Many students will be unfamiliar with backpacking or camping in primitive conditions. Have several evening sessions to demonstrate what should be packed and for what reason. How to cook, dress, set up camp and take care of sanitation are some of the topics that should be covered. Expect some students, unfamiliar with backcountry, to be fearful. Beware of the "expert" who has been on camping trips with organized groups. Enlist the help of experienced students to take part in a show-and-tell session of how they pack and travel in the backcountry. How to get there and back safely will require at least as much attention as the educational purpose for the trip.

May I?

Should an overnight field camp be part of the trip, be sure that you camp in an authorized location. Most parks and reserve areas have specific backcountry places for that use. Work closely with the park or reserve staff. Let them know when you are coming, what you plan to do and where. They may wish to send someone with you for your first few visits. Whether on public or private lands, pass through, make use of the land, teach students, and leave no trace of your passing.

Scenarios

Think through every eventuality before you go. For example: If a student sustains a badly sprained ankle and can no longer easily travel, then two others must assist. That produces three extra packs that must be carried by the rest. Suddenly a balanced group has become overburdened and in a difficult situation. Should you, the leader, become ill or incapacitated, who in the group could assume leadership? Make sure you have discussed this with students and that everyone knows who will assume what responsibilities and how the group is to proceed.

Ownership

Careful planning, inclusion, and cooperation are the keys to success. Each participant must have a sense of ownership and responsibility for the success of the venture. As part of preparation before departure, instill in each participant a protective frame of mind. *Each person is responsible for himself or herself, each person is responsible for every other participant, and each person is responsible for the group.* Stress to students that although they are adults and may have considerable freedom in their personal lives, they must be businesslike at all times while they are on school trips. Keep these trips SAD-free—no place for sex, alcohol, or drugs.

Travel Time

Maintain the sense of seriousness of purpose and stay on task during the trip to a remote site. You have not prepared adequately if information remains to be reviewed or first presented during the trip to the field site. All expectations and instructions should have been articulated, reviewed, and understood by all participants before the trip begins. Avoid raucous rides with loud conversations, cell phone calls, PDA messaging, music, and video games.

Eats

Food and its preparation will be a major topic. Small, single-burner camp stoves can be used to prepare a one-pot meal and cup of tea. Dehydrated food packets work well. The enterprising among the group may try mixing their own from combinations of soup mixes, rice, and other items to produce a "supermarket casserole." Heavily sugared foods may seem ideal but often cause distress. Bottles of soda and cans of beans and spaghetti may be quick and easy, but they leave a telltale aroma in the empty container that is tempting to furry friends. Think in terms of bulk, weight, and bears. Think in terms of trash to carry, pans to clean, yellow jackets, and mice. Remember there is no garbage disposal, no dishwasher, no sink.

Perfumed products and raw steaks do not belong in the backcountry, especially in bear country. The person with raw meat will find, after the first night, a pack full of holes, the meat gone, and a population of mice, gorged and content, scampering about.

Relief

Make sure you have sufficient drinking water for the duration of the trip. You also need to plan for restroom availability. Remote areas require a more primitive system. Check with the agency on whose land you are to travel for their recommendations. Avoid relief anywhere near standing or running water.

Using a garden trowel, an individual septic tank may be fashioned by removing a plug of vegetation and the topsoil. After it has been used and toilet paper placed in it, the plug is reinserted and stepped down, sealing the septic tank. Sealing is important to reduce the possibility of rodents digging down to it. (It is a food and mineral source.) It is important to excavate below the topsoil layer into the mineral soil below if possible.

Take Out the Trash

Be sure to bring trash bags to clean up your work areas, and carry out all trash. Good planning can greatly reduce the amount of trash generated. Experienced hikers are diligent in removing every piece that contributes to the weight they must carry. Even the tag on a tea bag should be left home. When you leave a study site, the next person or group to visit the area should find it at least as clean and unspoiled as when you arrived—or better. Leave no trace.

Be equally clear about items that should not be brought along, especially items that may add unnecessarily to the weight and bulk of transported materials and items that could dangerously distract students from educational tasks and safety precautions.

In Case of Emergency

In cases where cell phone service and other ordinary means of communicating with the outside world are unavailable, you must make certain your plans and schedule are known to someone who accepts responsibility for ensuring that a missed check-in or return will be reported immediately and search–and-rescue efforts initiated. The same person should also monitor weather and other warnings and, if necessary, let authorities know that you and your group may need emergency notification or evacuation.

What's the Weather?

We can't control the weather, but we can prepare for it. Know what the variations in temperature and weather can be at the site you choose. What are the risks of sudden storms or flooding? Be sure you find out, and be sure you have a shelter and evacuation plan.

Insist that your students are properly dressed for the weather. You may have students who come without gloves or hats. Simple gardening gloves or stocking hats might be a good extra provision. You are liable for making sure your students are dressed appropriately for all weather possibilities.

If you are planning a trip into the mountains, give special consideration to proper clothing. A generalization of mountaineers is that winter conditions may be encountered at any time when you are above 5,000 feet in elevation. During a midsummer trek to the heather meadows, you can encounter a sudden winterlike storm. Be prepared. The old trappers and prospectors were not wearing wool clothing for the style. They knew that it retains its insulating properties even when wet.

Lightning is extremely dangerous, striking and killing people each year. Even indoors, lightning can come through an open window or door and electrify a metal worktable. Instruct students to seek shelter if thunder is heard—at any distance. Once inside a shelter, do not use electrical equipment or work on metal structures. Don't take chances.

Flash floods are also life-threatening. You must check weather forecasts and escape routes immediately before entering any area where there is any potential for being stranded.

Dangerous riptides and "sneaker waves" can occur anytime in the ocean, but particularly on days before a storm appears or days after ocean storms have occurred miles away. Check regular forecasts, marine forecasts, and Coast Guard alerts. Be sure that all trip participants are aware of ocean conditions and changes that they may see.

Understanding the significance of a sudden tidal outflow—prelude to a tsunami—can be a matter of life and death.

Equipment and Supplies

When planning for an outdoor activity, you need to think about categories of equipment and supplies:

- Items needed to complete the planned activities
- Items needed to promote group and individual safety—such as eye protection, sun protection, and insect repellent
- Items needed for group comfort—such as clothing and shelter—for extended trips
- Documentation such as medical information and MSDSs

You also need to make sure equipment used outdoors is sturdier and less breakable than what you might use in the more controlled environment of your classroom. Try to avoid bringing anything made of glass and anything fragile or brittle—for instance, use plastic sampling containers rather than glass, metal probes rather than glass, and plastic hand lenses or field microscopes and water magnifiers rather than regular microscopes.

Weight and bulk should also be considered. Make student pairs or groups responsible for carrying and accounting for specific items, and then make sure the materials are packed for safe transport and are light enough for the students to handle easily. Plan sufficient time for equipment to be returned, counted, and repacked before leaving the field site.

PARTNERS

Your institution may have sports, fitness, or athletic training programs, whether as preparation for careers in these industries or as part of the college physical education or athletic program. You may find some helpful partners among faculty members or students in these programs.

For field trips into more rugged territory, these partners may be able help you, your staff, and students design a program that includes fitness assessments and progressive conditioning to prepare appropriately for the demands of the trip. They may also be able to prepare students with instruction in water safety and lifeguard training.

Select your partners with care. Be sure that they understand your goals and teaching style. The athletic skills of the field or court are not necessarily those of an extended backcountry trip, although there are some similarities. Be sure you and any partners are on the same page.

Apparel

Dress and footwear should not be left to chance or imagination. Make sure you provide students (and parents and guardians) with a clear list of appropriate clothing and shoes for the outdoor adventure. Dressing in layers is a useful strategy. It allows for adjustments to be made at the site. Hats are useful in both hot and cold situations. In sunny weather, they provide shade; in cold weather, they protect from loss of body heat. Shoes should provide good support for the arches and ankles and have nonslip soles. New shoes (not broken-in), open-toed shoes, sandals, flip-flops, and slipper styles should not be worn for fieldwork. Extra toe protection and waterproofing are highly desirable.

Communicate dress requirements clearly, and offer help if a student does not own the clothing needed. Many students don't own shoes or hats of the type needed for fieldwork or may not recognize the difference between ordinary sportswear and the clothing needed for backcountry field study. Predict the need for specialized clothing in your syllabus, so students are not surprised. Check all clothing and gear students are proposing to wear and carry long before the departure date. Hard hats are a minimal requirement in most geologic excavations and some archaeology digs. You may need to assist students who have financial or other difficulty in obtaining appropriate dress and footwear for fieldwork.

On extended, and many shorter, excursions each person should have a change of dry clothing. This change may be as simple as a stocking cap, wool socks, and set of woolen underwear double sealed in plastic bags—protected from the chance fall into a stream, pack and all. This emergency dry-clothes package need not be large or heavy. It does need to be carefully selected and waterproof packed.

First Aid

What to include in your first aid kit depends upon the hazards of the site. Plan your first aid kit item by item rather than generically. Involve students in the preparation and selection of first aid supplies, and determine what first aid material each student should carry and what should be carried for general purposes. Confer with the college medical staff concerning their recommendations and what specialty items you might carry on an extended trip. Ask the medical staff to instruct you on proper use of medical equipment and supplies. Bring copies of emergency medical information for all participants. For underage students, bring signed permission slips including permission for emergency medical treatment.

The standards for first aid best practice may have changed since you earned your last scout badge. Take a course, keep up-to-date, and get good manuals for yourself. For more about first aid, see Chapter 10 pp. 158–159.

Communications

Communications within the group and between the group and external help is a major safety consideration. For communications and check-in between groups and leaders, walkie-talkies set to the same channel may be more effective and reliable than cell phones, because they are not dependent on cell towers to function. Even walkie-talkies may not function in all terrains and should not substitute for keeping groups within

> **What's Not Needed**
>
> Certain items do not belong on field trips—or in student packs on field trips:
>
> - axes, knives, and similar hardware—Neither you nor your students should be hacking away at anything in a natural field site.
> - individual radios, compact disc and MP3 players, headsets, and video games—These items are distractions that can interfere with ensuring that every student hears and processes every precaution and instruction.
> - wrapping and packaging that might be discarded at the field site—Plan to carry out any trash that you may create at the site.

close range of each other.

Cell phones and personal data assistants (PDAs) have become ubiquitous. Whether you accept or reject our current connected state, you need to consider the effect of cell phones and PDAs on your field experience. The obvious consideration is disruption and inattention to activities. Fieldwork, by its nature, is away from the known, the everyday, and the security of the classroom or lab. Students calling one another or others, even to express wonder at what they are seeing, are distracted and can be dangerous to the group. For the purposes of the field trip, consider cell phones akin to a fire extinguisher. They are good to have around but should be seldom used and then only in an emergency.

Many field sites are far from cell-phone towers and service—some providers may be available in some areas while other providers are not. If you expect to use a cell phone in case of emergencies, do not assume that coverage is available. Check precisely both the location and the service provider.

Count, Count, Count, and Count Again

Count—equipment and personnel. Be sure that you have all that you started with and that you do not have more. Stop frequently during the day, not just at beginning and end, and conduct a count to be sure that every thing and everyone are present.

Take Nothing but Pictures, Leave Nothing but Footprints

> SCILINKS. THE WORLD'S A CLICK AWAY
> Topic: preserving ecosystems
> Go to: *www.scilinks.org*
> Code: SSCC148

If you concentrate on activities aimed at collecting data rather than specimens at field sites, you protect yourself as well as the environment. Before planning an activity that results in removing something from or irreversibly disturbing the field study area, ask yourself if there is any other reasonable way you can accomplish the same educational goals. Science is more about observing than about collecting, so the less intervention with the observed system the better.

If you and your students can observe without touching, so much the better. Let hands be on the instruments rather than on the organisms. That way, you minimize the hazards unknown or unanticipated organisms can pose and you make the fewest changes to the ecosystem you visit. You minimize contacts with potential allergens or infectious agents, and you avoid the inadvertent removal or harming of protected species.

Similarly, plan on carrying out everything you bring into a field site, including used materials, leftover supplies, and trash. Each outdoor environment has its own delicately balanced, perhaps unique, ecosystem. That is the reason you should not release classroom-raised organisms to the outdoors. They may be completely alien species, unable to survive, or, worse, they may have insufficient predators. The introduction of an alien species can negatively alter an environment forever.

And That's Not All

This chapter and much of this book discuss some actions that might be taken and identifies some actions that are best avoided. This is just scratching the surface. We can neither list all actions nor identify all areas of difficulty or danger. Every situation, every person, every field trip is different. We urge the reader and field trip leader to do their own research; plan thoroughly and think deeply about safety and responsibility. After the fact it is too late.

Hopeful and whimsical beliefs are no substitute for carefully planned and executed experiences. Do your homework, and work cooperatively with your students. Use this text as an introduction, an encouragement to think deeply and learn more about decisions that will be made, not as a final authority or detailed treatment—it is not.

THE SAVVY SCIENCE INSTRUCTOR

Mr. S and his colleague planned a spring break extended backpacking trip into the Cascade Mountains with a group of 10 students. All of the students had successfully completed the college's introductory environmental sciences class and had honed skills and techniques in observation and data collection during the program. Each participant had a conditioning program developed in conjunction with students in the college's health and fitness program and had been building up strength, agility, and stamina for the past eight weeks.

Finally, the trip. The first two days, everything seemed to progress as anticipated with only minor glitches—a student losing his balance crossing a stream

(cont. on next page)

9

(cont. from preceding page)

and soaking his pants, a dropped lunch, a closed trail. The group's preplanning ensured there were spare supplies and spare time.

But a more serious issue was in the offing. One of the students—the captain of the swim team—felt she was developing a "fluttering" heartbeat, something she had never experienced before. The second evening, she finally decided to inform Mr. S. The group had traveled half the route so forward or back would be equally difficult. The staff quietly held a conference and decided to proceed. Each staff member would keep a careful watch and quietly check with the student often, but not too often. The trip was completed without further incident.

All the planning had been well worth it. Though this specific event had not been anticipated, preparation had been made for a broad range of incidents including serious injury. The team had been thoughtful and careful in deciding how to handle the situation. Following the trip, Mr. S and his colleagues reviewed the entire event and their decisions. The college medical staff offered further insight and assistance, but all knew that their most important asset had been the ability to respond thoughtfully to the unanticipated. Would the team go again? Yes, in a heartbeat!

Connections

- *Amusement park physics.* 1997–2003. Annnenberg/CPB
  www.learner.org/exhibits/parkphysics
- Centers for Disease Control
  www.cdc.gov/travel/yb/index.htm
- Department of State
  www.state.gov/travel
- Kids' Lightning Information and Safety
  www.kidslightning.info/zaphome.htm
- Robertson, W. C. 2001. Community connections for science education: Building successful partnerships. Arlington, VA: NSTA Press
  http://store.nsta.org/showItem.asp?product=PB160X1&session=4183699CAA084FEC99AB1AF464AFF492

The Kitchen Sink

A Potpourri of Teaching Tips

At the community college level, the majority of faculty—full and part time—are well versed in their fields of instruction. Experience from business, industry, and the research sector provides rich content background. On the other hand, knowledge about instructional strategies, research in learning, and methods to promote safe laboratory practices may vary widely. In the previous nine chapters, we have tried to organize tips and technique by broad subject categories, using cross-referencing and repetition to try and provide comprehensive information within each category. However, a number of topics remain that do not easily fit in a recognizable category. We put them in our "kitchen sink"—a chapter full of tips and reminders that defy further classification.

It's Not What You Teach—It's What They Learn

Remember undergraduate science? Retrospectively, you may believe that you were responsible, mature, and always eager to learn. Perhaps you were—but that may just be the way memories mellow over time.

Creating a positive learning environment for adult learners with as wide a range of interests, goals, and abilities as found in a community college class is a creative challenge. It requires an understanding of learning styles, a wide variety of methods, and background in the field that is both deep and current. It's not enough to simply present information. It will roll off your students' cerebrums like water on a well-waxed vintage car.

Educational research in recent years has amassed a persuasive body of evidence to support the theory of constructivism. By combining the conclusions of many studies of different educational methods, researchers have found one factor that successful projects had in common. Instructors in those successful projects did not directly "teach"; the students actively learned by "constructing" knowledge for themselves. The critical role of the instructor was in motivating and coaching students to acquire and analyze data and draw their own conclusions, rather than lecturing them. But make no mistake, this type of instruction is carefully and thoughtfully planned, so that investigations are designed to highlight and develop important concepts rather than random ideas.

Instructors whose education was in a traditional lecture format may have to stretch to find methods that promote active learning. It is well worth the instructor's time to take some course work or workshops in methods of fostering inquiry. Among those methods are ways to structure laboratory investigations safely. That includes how to prepare students for a lab in advance, how to teach prerequisite skills such as measurement, observation, and equipment manipulation. It also includes ways to structure a syllabus to compel students to take more responsibility in lab work (including setup and cleanup). And it also includes direct instruction, but constructivism shows us many ways to structure lectures and demonstrations so that the audience is actively involved and engaged in what is being presented.

Give Me a Break

Whether you are responsible for planning a lab, a lecture section, or a combination of both, remember that no one concentrates well for more than about 20 minutes. Plan a change in activities and pace at least every 20 minutes. In a long laboratory period, plan several points at which you bring students together to check what they have done and what they understand. This might involve a short review of the safety procedures or the main point of the lab, or simply asking "Is anyone having problems with step ___?" That's not only a brain-friendly approach, but it discourages students from rushing through the lab to leave early. It's also safer, because students pace themselves and pay closer attention.

Another way to encourage students to move through your labs with due deliberation is to have them enter their data into a spreadsheet or database for projection in the last few minutes of the class. A scatterplot usually shows how most of the data points fall into the predicted pattern—and perhaps how one group has sloppy data.

There are many other ways to structure labs in shorter segments so that students get the most out of them. Here are just a few:

- Begin with a slide of possible preconceptions. Use a "think-pair-share" approach. First pose a problem that students must think through and record what they believe will happen and why they think so. Then ask them to share their ideas with someone else before they present the discussion for the whole class. (This works especially well for English language learner students, who can rehearse their ideas in their native language before they recite in English.)
- Have students prepare a flow chart or cartoon of the upcoming lab. Show by labels, diagram, and sketches the chemicals, activities, equipment, and flow to be followed. Preparing the chart or cartoon helps a student focus on the sequence of steps and anticipated outcomes of the exercise.
- Create a checklist of intermediate steps that students must complete. As they finish each subsection of a lab, have them report results in some visual or oral way.

- Encourage students to sketch their setup or observations in their notes. This approach helps students who are better visual than verbal learners.
- Provide digital cameras for students to record evidence of various steps in the process.
- Vary the methods or language that you ask students to use in presenting their results. For instance, ask that they write an abstract of their lab reports in the form of a news article, or as an explanation for a middle school or high school student, so that they avoid simply creating data charts that they don't understand.

Each of these ideas not only encourages deeper understanding but also supports safety because they are designed to force students to slow down and think about what they are doing.

Persistent Problems

Although it is relatively easy to analyze the safety concerns that might be related to a single lab activity, we sometimes take the general conditions of our assigned workspaces for granted. Some of the buildings we occupy are not as healthful an environment as they should be. Older structures may contain lead paint, asbestos insulation, or lead-soldered pipes. Asthma, allergies, and other persistent symptoms seem to be on the rise as buildings are made more airtight. Be alert to such signs as lingering coughs, sneezing, eye rubbing, headaches, and lethargy. They may be a signal that some irritant is present in your facility.

It's in the Air

Your ventilation system has an air intake register, a filter, ductwork to carry air, a motor-driven fan to move air, and an outlet exhaust. It is powered by electricity through a circuit breaker and has an on-off switch. It is much like a living system, and, like a living system if any part of it fails, the whole is in jeopardy. Periodically make an inspection tour of your workspace ventilation. Does fresh air actually come in, pass through your room, and then exit through the exhaust? Remember, don't assume—check it.

Chapter 3 includes the minimum standards for air exchange and for fume hoods. These standards are usually used for design specifications in new construction or renovation. But it is also important to regularly check to ensure that the installed system is functioning properly. For laboratory, preparation, and chemical storage areas, make sure that energy-saving programs do not shut down HVAC circuits when rooms are deemed unoccupied.

The Not-So-Magic Carpet

Carpets have never been recommended for laboratory classrooms. But some lab facilities may have them. A room converted from a regular classroom or borrowed may even have wall-to-wall carpeting.

If possible, request the removal of all carpeting from labs. This floor covering may be the source of a host of problems difficult to trace or resolve. A wet carpet makes an ideal breeding ground for the molds and mildew that are the most frequent and persistent cause of allergies and asthma in a closed environment. Spilled fluids can penetrate to carpet backing and padding, so surfaces that appear clean may be covering contaminated material below. Infrequently or improperly cleaned carpets can harbor dust mites, dander, and other allergens and disease-causing organisms. Glues and adhesives from newly installed carpet may outgas, causing mild but persistent headaches or dizziness. The recommended alternative to carpets in laboratories is seamless nonskid resilient flooring. Concrete is easy to slip on, and soft vinyl composition tile absorbs stains and chemical spills too easily.

In some cases, it may not be possible to remove carpet right away. For example, in older buildings, removal might trigger the costly process of removing asbestos in floor tiles below. If problem carpeting cannot be removed, request a thorough cleaning with mildew treatment regularly. Make sure the room is ventilated after treatment so the fluids can evaporate quickly. Even with thorough cleaning, some dander cannot be removed from carpets, so live animals, mammals in particular, should never be allowed on carpeted floors. The upholstery of old furniture may pose similar problems.

If you find that you have asbestos tiles (commonly 9" X 9"), consult an expert. Many institutions have a certified asbestos abatement plan, which specifies a date certain for abatement or removal. If that plan has gathered dust, resurrect it. It may also be necessary to explain to the facilities staff that placing carpeting over asbestos tiles is a bad idea for science rooms.

Heavy Metal Is More Than Loud Music

Curriculums have changed over the years. Many compounds once used for laboratory activities and cleaning in years past are now known to be hazardous—toxic or carcinogenic. Heavy metal and organic compounds can persist in cracks and crevices in flooring and furniture despite regular cleaning. Mercury used in older thermometers and other measuring instruments poses the greatest hazard. With the convenience and accuracy of new probe equipment, it's hard to justify use of mercury-filled instruments in introductory courses today.

If you see traces of spilled mercury in your work areas, request hazardous waste removal services. If you have older wooden cabinets, look for stains and other evidence of spills and test for mercury. In the event of significant contamination, blood tests for mercury exposure may be required. Some states require immediate evacuation of the room or building until all spilled mercury has been removed.

Cleaning Up

Thorough and regular cleanup is essential to safe lab practice and should become as much a part of laboratory responsibilities for your students as the collection of data. Standard housekeeping—leaving a clean lab bench following any ordinary activity—should involve the use of plain soap or detergent, warm water, and elbow grease. Do not be too shocked if you actually have to demonstrate for some students who have never used a sponge or dishcloth and don't know how to wring one out.

If disinfection is required, make sure that you inform students of the expected procedure prior to the commencement of the lab and that you allow sufficient time for proper cleanup before the lab period is over. Use the lowest-strength disinfectant that can decontaminate the surfaces and equipment, and be sure to rinse thoroughly. The mechanical friction involved in scrubbing down surfaces is often as important as the biochemical properties of the disinfectant.

Hand washing with soap and water should also be standard practice at the end of a lab. Again, it's the action as much as the strength of the soap. Little children are taught to wash for the length of a song like "Happy Birthday." You might want to demonstrate hand washing. Whatever technique you use, don't overestimate your students' knowledge.

In some cases, frequent soap-and-water hand washing by people with sensitive skin may result in chapping and skin breaks, which can cause problems of their own. In such cases, try substituting hand-cleaning gels made of quick-drying alcohol formulations. But keep in mind that alcohol gels are highly flammable and friction under dry conditions can cause some gels to ignite should a static electricity spark be generated. Another solution is to use nonlatex disposable gloves. Model the surgeon's technique of removing disposable gloves by flipping them inside out.

Some federal and state laws may restrict the amount and some types of materials used for cleaning, disinfecting, and pest control or require signage that identifies the material in use. Check to be sure you are not using some cleaning agent or disinfectant material that is banned or restricted or requires public notice. Use the information on the material safety data sheet (MSDS), and check with building administrators and facilities personnel. (For more information on MSDSs, refer to Chapter 4, p. 56.)

A good rule: Use the mildest possible chemical to perform the task.

The Right Cleaner

The newest cleaning formulas increasingly tout their germ-killing ingredients. It is important to recognize just how many and what kinds of weapons need to be brought to the battle with germs. More often than not, just plain soap and warm water are enough.

10

Standard (Universal) Precautions—Typical Practices

- Never bring blood or blood products into the classroom/laboratory.
- Always have nonlatex gloves available for use in cases of bleeding.
- Provide sterile gauze pads to the injured individual to cover and hold over the wound.
- Do not interfere in a fight in a manner that could expose you to blood or biting.
- Use approved disinfectants for blood or body fluid spills. To prevent splashing and further contamination, cover the spill with paper towels or absorbent cloth such as an old T-shirt, then pour the disinfectant over the towels or cloth.
- Use a specially marked disposal container for all materials contaminated with blood and other body fluids.
- All designated personnel who might be asked to clean up body fluids such as blood, saliva, or vomitus should be provided the opportunity to be vaccinated against hepatitis B.
- Have a special waste container for biohazards.

Refer to college health services for the specific practices adopted by your institution.

Sometimes a 10% solution of household bleach solution is all that is needed. Bleach solutions lose their potency quickly and should not be kept longer than 24 hours. Bleach solutions should not be used with any other soap or cleaner and in particular must never be mixed with ammonia solutions. A chemical reaction may result in the production of dangerous gases.

Antibiotic hand cleaners are not recommended for daily use. Science classes should not contribute to the selective evolution of supergerms.

MSDSs are required for every product in your room. This includes all cleaning products. This may involve some inservice education for your custodians, who might switch products without notification. (For more information on MSDSs, refer to Chapter 4, p. 56.)

Potable Water

There should be no drinking, eating, or storing food or beverages in any science laboratory. Water from the taps in science labs should be used only for lab work and cleanup. In addition to this being a requirement for lab safety, many older buildings have iron pipes with lead solder that can leach lead into the water, rendering it unsafe for drinking. Direct students to water fountains outside of the laboratory for their drinks.

Standard (Universal) Precautions

When the specter of HIV/AIDS first sent its shadow across the world, people debated how much information should be shared about infected persons. Soon we came to realize that it was far more appropriate to adopt safe practices and assume that anyone and everyone could be infectious. In response to hazards associated with bloodborne pathogens, the U.S. Occupational Safety and Health Administration (OSHA) required that each institution adopt a set of Universal Precautions for handling human blood.

Today we recognize that not just HIV but also many other pathogens can be transmitted not just by blood but also by other human body fluids such as saliva, semen, and vomitus. The original Universal Precautions were expanded to include the additional risks and became known as Standard Precautions. Today, the terms *Universal* and *Standard* are commonly used for the more inclusive Standard Precautions. As with Universal Precautions, Standard Precautions are based on the assumption that any person is potentially infectious and that all human body fluids, regardless of source, should be assumed to be infectious—no exceptions.

People likely to come in contact with human body fluids should be initially trained and then receive additional yearly updates in practices designed to protect against transmission of pathogens found in body fluids. Hepatitis vaccination is normally recommended for people with responsibility for handling or cleaning up body fluids. Each institution must adopt its own set of practices for handling body fluids. Some typical practices are highlighted in the box on p. 156, Standard (Universal) Precautions—Typical Practices.

Use and Disposal of Sharps

Sharps include tools, parts of tools, and broken objects with sharp edges or points that can break the skin and cause cuts or other injuries. Metal blades, needles, broken glass, and plastic with jagged edges are all considered sharps. If you use materials that have sharp points or edges or that can be broken into pieces with points or sharp edges, you must prepare a separate disposal container for them. This container should not be opened or emptied but rather disposed of intact and replaced with a new one. Make sure everyone who uses your room or handles your trash knows about the sharps container, uses it, and handles it correctly.

If you don't already have them, you and your students can easily make sharps disposal containers.

Hepatitis

- Hepatitis A is a liver disease spread by bad hygiene and contaminated food.
- Hepatitis B is a viral liver disease spread by body fluids, including saliva and blood. It can be spread by biting and spitting. It is often chronic.
- Hepatitis C is a viral liver disease spread by body fluids. It is the least-known form and often lies dormant for many years before causing serious symptoms and sometimes death.

Old Classics

Current knowledge of the potential for infection from body fluids also affects many standard instructional laboratory practices. Lab exercises that involve human body fluids (even the students' own) should be deleted from the syllabus for introductory classes. This includes old standbys such as blood typing and blood smears as well as experiments using and testing saliva, cheek cells, and serum. In more advanced courses, you must enforce and practice the Standard Precautions on record for your institution and make sure that your students are fully instructed and capable of using the same Standard Precautions.

Use a plastic gallon jug or a puncture-resistant box sealed on all sides. Cut a small slot in the container large enough to accommodate broken or sharp items for disposal. Label the sharps disposal container prominently on all sides—color it red or place a red label on it. Make sure your custodian will recognize it immediately. If an object with sharp or broken edges is too large for your regular sharps container, make a special one just for that object. Keep one or two small cardboard boxes just for this purpose. Be sure to seal and label the special sharps container clearly, and place it near the regular sharps container for disposal.

Inventory all sharp items distributed for use during class. Razor blades, even "safety" razor blades, should not be used for dissections or class work. Take a tip from the shop instructor and try organizing instruments in a keyed holder so that missing items are easily spotted. Another method is to color-code each group's instruments with nail polish, permanent markers, or paint to make an easily seen visual pattern such as a diagonal stripe that highlights missing items.

If someone does get cut, be sure to observe Standard Precautions for handling body fluids in treating the blood and wound and maintain a written record of the injury and the procedures followed.

The Latex Connection

Latex has been identified as a serious allergen for many people—causing reactions as simple as rash and irritation and as serious as anaphylactic shock. Common sources of latex in the classroom are in "rubber" gloves, both disposable and nondisposable. We recommend non-latex (nitrile) gloves instead of latex. In some cases, other glove types are needed for specific types of chemical or substance handling. Know what glove is correct for what use. Rubber tubing, rubber dams, and some flexible connectors may also be made of latex.

Another source of latex in the classroom is the common balloon. In addition to their allergenic potential, balloons present a choking hazard if students fool around while blowing them up. There are other reasons to avoid using or releasing helium balloons. They can do a lot of damage to birds, other wildlife, and the enviroment.

First Aid

It's prudent to be prepared for the unexpected. We highly recommend first aid training courses from reputable organizations. Most states have Good Samaritan laws that provide liability protection for a trained emergency response. Check to find the specifics in your state.

For a person with a severe allergy, an EpiPen or adrenalin injection can be of critical importance. Without it, anaphylactic shock from an insect sting could result in a fatality. Although relatively simple to use, the injection must be administered only by prescription by a nurse or person trained by the prescribing physician. In planning for fieldwork you may need to make sure that you or someone else is trained to use an EpiPen.

Red Means Stop, Green Means Go—Or Do They?

Color blindness, particularly red-green color blindness, is more common than most people suspect. If you use color-coding in organizing your lab, particularly if you use red to symbolize a safety warning or precaution, make sure you include verbal or graphic indicators so color-blind individuals do not miss your meaning or distinction. Color-blindness, a sex-linked trait most often appearing in males, often goes unnoticed even into adulthood until some comic or unfortunate incident makes it apparent. Some men in your college classes may not realize they are color-blind, even though they may be mildly to severely affected.

The Scientific Gourmet

There may be some activities in lab guides or your syllabus that involve food or cooking. Special precautions must be involved.

No Cooking, Tasting, or Eating

There should be no eating or tasting in a science facility. The hidden dangers that come with consumption of food or drink in a science laboratory fall into two categories. First, the area may be contaminated with surprisingly persistent toxins, including heavy metals, organic compounds, molds, and pathogens. In a shared science space, you can never be sure of what materials were there before or how well the space was cleaned. Second, students who are in the habit of eating in a science workspace may be tempted to taste a material that is meant for research.

The best rule is the most simple: Nothing should be tasted or eaten as part of science lab work. No snacks or food should be eaten in a science room or in the part of the general room where investigations have taken place. This policy is consistent with regulatory requirements in all areas where hazardous materials may be present including, but not limited to, research facilities. The presence or consumption of food in a laboratory can result in a shutdown of the entire operation.

But what about those motivating experiences that involve foods, such as observing changes in popping corn or measuring the calories in croutons? There is no problem in using edible material in lab activities as long as none of it is consumed.

Nut Allergies

An increasing number of children and adults have been identified with serious—often life-threatening—allergies to nuts. An allergic reaction may be triggered not only by ingestion but also by proximity to a nut or nut product. If such an allergic individual is present in your class, you must avoid investigations involving nuts, nut oils, and nut by-products. Substitute croutons for peanuts in the traditional calorie laboratory.

Lab Refrigerators

Refrigerators in laboratories and in science storage and preparation areas must never be used for food or drink meant for human consumption—no exceptions. To reinforce this, make sure that every lab refrigerator has clear signage that says: laboratory supplies only—no food or drink.

Laboratory refrigerators for the storage of volatile and flammable substances must be sparkproof—not the standard household refrigerator. If a lab refrigerator is not sparkproof, clearly indicate this with signage.

Dress of the Day

Dress codes may seem old-fashioned or even controversial, but safety takes precedence. Some clothing fashions are just poor choices for laboratory work. It is both legal and practical to set a commonsense dress code for the laboratory. In advanced classes, specific gowning and head coverings may be required. The laboratory is a workplace—an environment where safety trumps fad and fashion. Loosely hanging, floppy clothing and hanging jewelry are not reasonable. If a student insists a particular piece of jewelry cannot be removed, cover it completely with tape or a bandage.

Coats, jackets, hats, and similar loose-fitting, overhanging, and dangling articles should be removed and stored away from work and lab areas. Midriffs must not be bare. Absorbent watchbands and wrist ornaments should be removed. Hair should be tied, pinned, or otherwise secured back behind the shoulders. Backpacks and totes should be stored in lockers or other storage areas. Shoes should have closed toes and be securely tied. No platform shoes, flip-flops, or any other footwear that is likely to result in tripping and falling should be worn in a science laboratory. This is a matter of safe practice rather than fashion commentary. Instructors and assistants must conform to the same dress guidelines.

Combustible Fabrics

Even when you've controlled for fashion, you still may have trouble with the fabrics in student clothing. Many fabrics are highly flammable and may melt at relatively low temperatures, causing severe burns and permanent injuries. You may wish to demonstrate the high flammability of fabrics, especially filmy, gauzy fabrics. Instruct students about what to do if clothing catches fire:

- Stop, drop, and roll.
- Bring the fire blanket and extinguisher to the victim rather than vice versa.
- Should hair catch on fire, smother the flames. Stop, drop, and roll is not effective in this case.

10

All Eyes on Science

Clear vision and protection of the eyes are musts in almost any laboratory activity.

Contact Lenses

The use of contact lenses in the laboratory is the subject of intense safety discussions. Although regulatory agencies may not consider contact lenses to be a hazard, some organizations and eye-care professionals do have very serious concerns and recommend against use of contact lenses in laboratory activities where fumes may be absorbed by the lens. However, there is complete agreement on the need to wear appropriate eye protection as a first defense. The reader should do his or her own research to reach an understanding of this issue.

Eye Protection

Many instructors associate eye protection with chemistry activities, but it is also a must in many other circumstances, including whenever there is the potential for fire or projectiles. When eye protection is needed, everyone present in the laboratory must comply—all students, instructors, assistants, and visitors. We recommend the use of safety goggles rather than safety glasses and that each student has his or her own eye protection.

We recommend you encourage students purchase their own safety eyewear, because it is difficult to sterilize a class-load of goggles in the few minutes between lab sessions, and risky for a faculty member to assume that the job was done correctly in the previous section.

> **Recommended Eye Protection**
>
> We strongly recommend that you only use safety goggles that have chemical-splash protection and are ANSI Z87.1 compliant for impact protection. If your students purchase their own goggles, make sure that the bookstore knows what to order and sell—goggles that protect against both impact and chemical splash.

Any activity that involves use of any chemical (in liquid, solid, or powder form) that could injure the eye—including dissection activities—requires splash-proof eye protection. Chemical-splash goggles are indirectly vented and have baffles that permit air to circulate without allowing harmful chemicals to enter. Students sometimes remove or flip open the baffles. Be sure to warn against that practice and insist that baffles be in proper position.

Projectiles present an impact hazard. ANSI Z87.1 is a voluntary impact-resistance standard that can apply to safety eye protection devices.

We strongly recommend that safety goggles providing chemical-splash protection and conforming to ANSI Z87.1 impact-resistance standards be used in all science laboratory work where eye protection is needed. Regular "street" eyeglasses and "plant

Safety Eyewashes

We recommend that you use only plumbed safety eyewashes and not the bottled type. The water must be tepid (not cold) so it must be plumbed appropriately, and the pipes must be flushed regularly. (See Chapter 3, p. 36, for more information on eyewashes.)

visitor specs" do not provide sufficient protection in science laboratories.

Neither safety glasses nor safety goggles are intended to protect the rest of your face and your throat. In some instances, a face shield may be necessary to use for that purpose, in addition to either safety glasses or safety goggles. Face shields are often used in college laboratories, and instructors may need them when they prepare dilutions and solutions. But we strongly urge that you reconsider any student experiment that would require the use of face shields. It would be better for community college students to use smaller quantities, different reagents, or simulations.

Disinfecting Safety Eyewear

If safety goggles are shared, they must be sterilized after every use either in an ultraviolet (UV) cabinet or in hot water and detergent or disinfectant to prevent the spread of conjunctivitis and hepatitis. UV cabinets are not as high-tech as they seem. If goggles are just tossed haphazardly into them, the light sometimes doesn't reach all the surfaces and disinfection is unreliable.

A simpler procedure is for departing students to drop their goggles into a sink filled with dishwashing solution. Entering students can thoroughly rinse and dry the goggles and wear them immediately afterward. If you use this method, make sure straps are plastic rather than cloth. Have plenty of soft clean towels available, because rough paper towels can scratch the plastic lenses.

For optical instruments, disinfecting may be more difficult. Remind students that, to use a microscope properly, you do not need to touch your eye to the eyepiece. Harsh cleaning solutions or alcohol applied to optical instrument lenses can permanently damage the lenses and lens coatings, so check manufacturers' care guides and follow instructions carefully.

Conjunctivitis

Conjunctivitis (sometimes called pinkeye) is highly contagious and can easily be spread from one user to another via contamination of safety eyewear and optical instruments such as microscopes, binoculars, and telescopes. There are both viral and bacterial forms of the infection. Decontamination and disinfection of shared-use safety eyewear is a must. That's why, if conditions permit, asking students to purchase their own safety eyewear is a better option. Coordinate with the campus bookstore to ensure that it stocks only goggles that provide both chemical-splash protection and ANSI Z87.1-compliant impact protection. Goggles can be sized to fit an individual student's face snugly.

Standards and Agencies

There are quite a number of standards promulgated by federal agencies, industry, and professional organizations. Not all states or institutions are required to comply with all these standards. We strongly recommend, however, that you familiarize yourself with guidelines from the following agencies or groups.

OSHA

The Occupational Safety and Health Administration (OSHA) was created by the Occupational Safety and Health Act of 1970. It is the federal agency responsible for developing and enforcing workplace safety and health regulations and for investigating workplace accidents. Although the standards apply to employees (they do not include students or the general public) and not all states subscribe to OSHA standards (many states have their own plans that sometimes exceed OSHA standards), the requirements do provide guidance on safe procedures and operations. We urge you to check OSHA guidelines and your state regulations if you are not in an OSHA state.

NFPA

The National Fire Protection Association is an international nonprofit group that attempts to reduce fire and related hazards by gaining consensus and acceptance of codes and standards, research, training, and education. Conforming to their recommendations is highly recommended and, under some circumstances required, in the design of laboratory facilities and selection of fire safety equipment.

ANSI

The American National Standards Institute is a private, nonprofit organization that administers and coordinates the U.S. voluntary standardization and conformity assessment system, including the ANSI Z81.1 standard that we strongly recommend you look for in selecting safety goggles.

BOCA

Building Officials and Code Administrators International (BOCA) is a founding member of the International Code Council (ICC). Their guidance is invaluable in the design and construction of science facilities.

Your Maintenance Staff Partners

Members of the maintenance staff are critical to safe laboratory operations. Though you may not have frequent direct contact with them, you must ensure that they are aware of the hazards that exist in your laboratory and take steps to protect them from exposure to hazardous materials generated by you and your students. Clear communications with them and respect for their role are crucial.

Make sure that custodians know how you are handling sharps, biohazards, and organic and chemical wastes. If you are keeping live animals or cultures, be sure that the maintenance personnel are well aware of the conditions that must be maintained.

Establish a schedule in conjunction with the maintenance staff to regularly test and flush safety eyewashes and showers. If floor drains are not available under safety showers, large trash barrels may be required when showers are tested. Keep records documenting your regular tests. These records may be as simple as a printed spreadsheet with spaces for jotting the date tested or a tag similar to those used on fire extinguishers.

Keeping in Touch

For safety, for liability, and for just plain good education, it's important to keep in touch with your students. That can be difficult; they may want your attention day and night. If you are full time and have office hours, it may be easy to establish limits. For faculty who move in and out of the college, student contacts can be sporadic and unreliable. But they are very important; failure to respond to a pregnant student's question about the materials in a future lab or to hear a request for accommodation from a special-needs student could be a serious error. Use your school's online learning platform or your e-mail system for confidential questions from students.

Telephones

A phone is a vital piece of safety equipment. No science lab should be without one. The landline, *not* the cell phone or Voice Over Internet Protocol (VOIP), is the most reliable form of communication. Make sure you know how to get an outside line to 911 and an internal line to facilities staff before you ever invite a student into your room. (Some VOIP phone systems don't have 911 access; always check.) Post emergency phone numbers and calling procedures by each phone.

Voice mail is available at almost every institution in the United States. But voice mailboxes are limited, and, even in the best of systems, messages sometimes get lost. If you rely on voice mail for student communications, keep a careful log of the calls you receive and the responses you make. Phone tag may be amusing in a social context, but, when you are managing a science program, it's not a game you want to play.

The flip side is student phones. They aren't just rude; they can be a hazardous distraction. Students have no business with active cell phones or text-messaging devices while they are in class. Make this a ground rule, and stick to it.

E-Communicating

E-mail can be an organized way of maintaining a record of student communications. Tell your students in advance how to reach you and when to expect an answer. Responding within 24 hours would be considered normal. Having an e-mail system in each course's platform is preferable to your general, personal e-mail because it is exclusive to your class and usually makes it harder to accidentally delete a message.

Use your college's platform or your webpage to post an extra version of lab instructions and safety issues that students can preview in advance, with links to the MSDS sheets for each chemical you use. We also recommend you post information about possible accommodations for special-needs and disabled students. That may seem complicated at first, but, once it is done, it's a record that's extremely valuable. Take the time during the first class to show students where to find the information you have set up and how to read MSDS sheets.

Set up a special folder for student questions and your responses. Let students know in writing, in the syllabus, that this information is available. When you provide a response to an issue with potential legal or safety ramifications, copy (don't blind copy) your message to another person in your institution with supervisory responsibilities, such as someone on the health or disability staff. That's just one more way you cover yourself when students have unanticipated health issues or concerns. Keep a copy of the student question and response folder for at least a year. Do not rely on the hard-drive records of Outlook or other desktop communication systems. Because of the way they are stored, it's very difficult to recover them if your computer fails.

The Internet Connection

The internet is a double-edged sword when it comes to safety and reliable science information. The information posted on websites need not be edited or vetted. False, misleading, and downright dangerous information is common. There are still people putting tinfoil on their heads to protect themselves from cosmic rays, and the internet simply gives them an extraordinarily efficient means of spreading their theories. Teaching students how to vet internet sources is an important skill.

Each instructor should take specific training on how to check student papers for plagiarism from the internet. Make sure students respect copyright of text, images, music, and other elements available on the internet. If your institution has developed an internet-use policy and a written contract to be signed by students, be sure you know the policy and follow it carefully.

In addition, make sure that the answers or responses to assignments you give cannot be found easily on the web. Many instructors use the same books, and, unfortunately, both students and instructors sometimes post standard answers to key questions.

Save Paper and Reduce Clutter—CD-ROMs and Flash Drives

CD-ROMs and flash drives are convenient tools for storing documents. Documents prepared with almost any word processor can be converted to portable document format (PDF) files. Adobe provides free downloads of PDF readers on its website, *www.adobe.com,* thus allowing students and others to read and print but not revise your documents even if they do not own a copy of your word processor.

If all your students have access to the technology, you might consider preparing and distributing CD-ROMs containing class information in PDF format or posting the information on a website or platform for students to download. Starting with your syllabus, class documents that are to be shared with students can be placed on your website or on a CD-ROM that is given to each student at the beginning of the term. Laboratory and safety issues can be prominently displayed and easily accessed. MSDSs for chemicals can be placed in an MSDS folder. Digital pictures of safety equipment and other items can become part of the CD-ROM record. The initial setup to bring all your material together will be time consuming. Like an inventory, however, once it is done, it can be easily modified as new editions are prepared. It will present a whole picture of your class to you and your students, and it will be a valuable addition to the class. Additional information can be added to most CD-ROMs. But don't make this the *only* information you provide, since you can't assume that every student has access all of the time.

Stay current. Review and revise your files at least every other term.

Independent Studies

Project-based learning and laboratory investigations sometimes allow students to shine who have not done so in traditional settings.

For some of these students, independent studies at the college level can be valuable add-ons or can constitute an entire course of study. But advanced and independent projects can also raise significant safety problems. Remember that independent study asks students to go beyond their knowledge base and beyond their science competence to investigate unknown phenomena. It sounds good: students may rise to the challenge and instructors enjoy seeing student growth. But be prepared to work closely even with a good student for the entire course. Accept independent study students with caution and care. Remember that safety is as important as content. Success is rewarding, so create the conditions for success carefully. Consider including the following steps as part of your oversight:

- Prior to considering the project, sit down with the student and discuss the idea and its complexity and feasibility.
- Require a written proposal including an outline of proposed laboratory activities.
- Establish a procedure that allows you to review all aspects of a laboratory investigation before it begins.
- Make the student responsible for obtaining MSDS information and safety instructions on all substances and equipment to be used. Review this information and keep in your files a dated, written copy initialed by the student before the experiment.

- Make sure you are there whenever an independent study student is working in the laboratory.

Neither a Borrower Nor a Lender Be

Offered centuries ago, Polonius' advice to his son in *Hamlet* can be applied aptly to community college instructors—as a corollary, be neither giver nor taker.

Community college instructors often straddle the realm between high schools and college, either as instructors in both venues or as friends and advisers to secondary school teachers. That can lead to a strong temptation to give or lend supplies and equipment to resource-limited secondary colleagues—or to use supplies and equipment from your community college storeroom in your high school classes. Our strong advice is: Don't. The responsibilities and liabilities are numerous and onerous.

If the request comes from a secondary school friend or colleague, you could be held responsible for the use, misuse, or loss of the material. If proper safety precautions and safety equipment are not in place and used appropriately when material you supplied is in use, and an accident occurs, you could be dragged into a costly liability suit with no institutional protection.

If you are a secondary instructor as well as a community college instructor, we strongly caution you to avoid lending or giving items to yourself. If your secondary school does not provide you with the supplies and equipment you would like to use for that extra demonstration or experiment, it is also likely that you do not have the proper facilities and safety equipment to carry on those activities safely. Which institution will back you up and accept the liability if an accident occurs?

Fair Well?

Community college instructors may be invited to judge secondary school science fairs. Before agreeing, we urge you to inquire about the rules and safety measures that will govern the event. Consider some of the following questions, and make sure you are satisfied that the answers are adequate:

- We have all heard the tales of such events becoming a contest between parents rather than a chance for students to engage in investigations of their own, so what steps have the science fair organizers taken to prevent such a takeover?
- What steps have been taken to ensure that the work is the student's own?
- Will the "projects" be science investigations or science displays?
- How will you determine whether the student fully understands his or her own investigation and the resulting data?

- Will safe laboratory practices be a consideration in selecting the winner?
- What safety measures and equipment will be in place during the fair?

Remember, and remind your secondary colleagues, that the science fair is likely to draw crowds of visitors, including young siblings and friends, the curious, and sometimes surprisingly irresponsible adults who may be interested in showing their own "expertise" with little thought of safety. Are explosive demonstrations a possibility? Could a hapless visitor open a staph-loaded petri dish cultured from the hand of an ill-advised young scientist? Could there be glassware exposed to pressure or heat? Are toxic chemicals a possibility? At the very least, review the "Minimum Safety Guidelines for NSTA Presenters and Workshop Leaders" at *http://ecommerce.nsta.org/sessions/pdf/440.pdf*.

Crisis Prevention and Response

As the all-too-common headlines of violence and crises on campus show, this book could never describe everything needed for security. We have all had to change our behavior, our thinking, our approach to the new realities. What we hear, see, and do on campus can be important—more than we wish to think. Materials obtained in our labs may be used for more that just science experiments. Bumping along, trusting to luck, and dreamily hoping for the best is fast falling from acceptance. Probably, the most effective safety factor is the staff. A caring, watchful group of faculty can spot potential problems before they are even imagined.

Science instructors in particular can play key roles in crisis prevention. You have many opportunities to work with students in informal settings. When your students are engaged in lab activities in small groups or pairs, you have a good opportunity to observe and hear students' attitudes emerge. Put up your antennae for signs of troubled students. Report your observations to student support staff. Follow up in writing (ideally, with an e-mail record). Never ignore threats of violence or suicide or other warning signs from students, assuming they are just kidding.

Keep outside doors closed, and make sure the students who enter your room are those enrolled in your class. Don't allow friends to enter, or students to stray away from the class. If you spot an intruder, notify campus security at once. College campuses are open and available—an invitation to learning. A new more observant perspective, an observing mind-set could be an important change in an instructor's behavior. Not all visitors on campus are answering an invitation to learning.

Keep your room locked when you aren't present. Unlock back rooms, prep areas, storage areas, and cabinets only during the time that they are in use. This may be a challenge, especially when you share space with other faculty. But you may be held liable for what happens in your lab when you are not there. Don't have your lab set up too much in advance (no matter how persistent that busy lab assistant might be). Once it is set up, it's your responsibility to maintain security for the materials.

Plan for the worst so you can securely enjoy the best of times. With your fellow staff members and the support of the administration, create and train crisis response teams to respond to medical and psychological crises, threats and acts of violence, and the range of circumstances that require immediate action for the welfare and safety of students and others on campus. Such teams should plan for emergencies and ensure that individuals are trained and available to provide first aid, psychological support, physical security, and public information. Include and maintain regular contact with representatives from state and local agencies that provide assistance and emergency response—police and fire departments, community mental health agencies, local hospitals, and state health departments.

Pregnancies

Take note of signs of illness and refer any student with possible problems to the proper medical facility. In science, some chemicals—such as some biological stains and preservatives and many solvents and organic compounds—are teratogens and pose serious risk to the fetuses of pregnant women.

Because you may not know if a student or colleague is pregnant, the best thing to do is to completely avoid the use of teratogenic chemicals. This is another great reason for having a complete, clear syllabus. Describe each of your labs in detail, and provide an explicit list of chemicals, supplies, and equipment to be used so students may consider possible hazards in general and for themselves specifically.

Promises to Keep

Provide opportunities for informal conversation with your students, but do not promise secrecy. You need to assure students you will respect their private communications with you but not at the expense of failing to exercise your judgment in protecting them and others.

Privacy

It is your professional and legal responsibility to protect the privacy of students and colleagues. Do not disclose information concerning medical or psychological conditions except to licensed professionals responsible for protecting the individual's health and welfare.

Grades and other educational information likewise should not be disclosed except directly to the student or the minor student's parent or guardian. Avoid disclosing such information via e-mail or telephone unless you can be certain of the identification of the recipient. Refer to Family Educational Rights and Privacy Act (FERPA) at *www.ed.gov/policy/gen/guid/fpco/ferpa/index.html* for further information.

Planning for the Future

Don't wait until the end of the semester to make plans for the coming year. When your department head begins to discuss assignments for the next session, schedule an appointment to consider the priorities below.

- *Class size:* Although additional facilities and equipment can make science safer for larger groups of students, the number of students that can be effectively and safely taught in an active investigative science program has a limit. There is a very high, positive correlation between class size and accident rate no matter how good the facilities. Although experts disagree on an absolute limit, statistics show accidents increase dramatically as classes increase beyond 20 to 24 students. Refer to National Fire Prevention Association (NFPA) and Building Officials and Code Administrators International (BOCA) for current occupancy load information.
- *Scheduling:* Preparing for laboratory work takes thought and time. Equipment must be inventoried and checked for correct and safe functioning. Surfaces must be cleaned before and after messy activities. This kind of preparation can't be done on the fly, as an adjunct is shuttling between day and night jobs. That means that the safest schedule for everyone involved may not be the most convenient for the most senior staff.
- *Security:* Two exits are needed for science rooms. If your state does not require two exits for a science laboratory, request a room with two exit doors that is not in a heavy-traffic area or near a major school entrance. If you suspect old copies of your classroom key are floating around (odds are they are), ask for your room to be re-keyed. Keys to science storage areas and rooms solely dedicated to science instruction should be unique. (See Chapter 4.) These rooms should not be accessible with keys used to enter regular classrooms or common areas.
- *Physical Plant:* This is a good time to look to your surroundings and develop a positive action plan. What changes or repairs need to be made to the science facilities? Is there some modernization that would be a benefit to teaching science? What are your long-term goals? Does the ventilation system need overhaul and upgrading? Do you need additional secure storage areas with air vented directly to the outside? Do the eyewash stations and safety showers supply clean and tepid water? Are there older glass-faced cabinets you would like removed from the teaching area? Now is the time to develop that plan with your department and establish a timeline to make the changes that would provide a better learning environment. Remember that many of these changes may require earmarking future funds and taking time to plan and complete. This requires a proactive approach on your part. Follow your request through channels and stay with it to completion. If your facilities requests are scheduled for several years hence, make sure they do not get dropped off the next versions of multiyear capital improvement plans.

Test Time

College students have already learned one important lesson: The instructor values what is put on the test. That's why it is important to assess safety skills in use, not just at the start of the term but also with every lesson. Use multiple assessment techniques to determine comprehension of safety lessons. The diverse student body requires working effectively with students who are older, who speak and understand English as a second language, who are handicapped, or who have other special circumstances that may make comprehension more difficult. Do not assume that just presenting a lesson, or warning, means that it is understood or will be acted upon. Be sure that the message has been understood.

Although we can't provide a complete manual of ideas for authentic assessment in safety education, consider these:

- Require specific skills at the outset. Before beginning a lesson, students may need to master accurate use of graduated cylinders, burners, or balances. Provide rubrics that permit students to self-assess or assess each other within a group. Build lessons in a way that requires students to master techniques and safe practices to their own satisfaction and that of their peers before proceeding with the rest of the investigation.
- Require students to embed safety precautions into every lab report and portfolio presentation. Precautions that students themselves develop can be especially effective.
- Use safety scenarios as stems of multiple-choice items.
- Use a digital camera to photograph different setups. ("What problem do you see here?") Or describe the effects of improper equipment use in a short paragraph. Students should be able to recognize not only the precautions they should take but also the results of poor techniques.
- Use news clips or stories of real-life accidents to prompt students to demonstrate their authentic understanding of good safety techniques.

After you use an assessment to identify a deficiency, act. Develop a remediation lesson, an alternative method of helping students master the material, or a new lesson plan.

Everything and the Kitchen Sink

You may have found this book has touched on topics you never thought were the responsibility of a college science instructor. Some issues may be new to you, such as your responsibility to ensure that students with special needs have every opportunity to acquire the same knowledge and skills as students without special needs—even if you must make special accommodations and modifications in your methods of teaching and assessment. But they are indeed very much your legal responsibility.

In other cases, you may not be primarily responsible, but an awareness of the issues and an understanding of appropriate decisions are a part of your responsibility. Our aim has been to be comprehensive, even though not exhaustive, in providing you with thorough information as well as information that you may need to share with department and institutional administrators.

THE SAVVY SCIENCE INSTRUCTOR

Dr. F has a three-hour lab section on Monday nights, beginning at 6 p.m. By the time the section meets, it seems everyone is tired. Most of her students have worked all day. So has she.

Because she is an adjunct, Dr. F doesn't have an office. To create an informal way to meet students and give them a chance to shift focus from traffic jams to science, she has set up a coffee pot in a small unused room across from the lab. Because there have never been any chemicals stored or experiments done in the room, she can comfortably invite students to bring cookies or chips. A sign on the door indicates that the food stays there. By the time she unlocks the laboratory, everyone is ready to focus.

Most of the labs in Dr. F's curriculum are long and detailed, with several stages to each lab. When she first began teaching, Dr. F used to present the entire lab in one session. Some students would rush through, finishing in two hours. She'd only discover a week later that they had missed a step or ignored an important observation.

Now she uses technology to encourage a slower, more deliberate investigation. Each table has a digital camera. Students are encouraged to stop when they reach a specific point in the lab and record their setups. If there is numerical data to collect, they move to the front and enter it into the instructor's computer. Dr. F schedules think breaks into the lab at key points. These are also breaks at which students may leave their apparatus and go across the hall or to the lavatories, or even step out to make a phone call. Other times are strictly business.

10

The photos and data that Dr. F collects at think breaks can be used for lab quizzes, or just for discussion. She's found that her long lab evenings are now less stressful, more productive, and much safer.

Connections

- American Chemical Society and ACS Board—Council Committee on Chemical Safety. 2001. Chemical safety for instructors and their supervisors. Washington, DC: ACS. *http://membership.acs.org/c/ccs/pubs/Chemical_Safety_Manual.pdf*
- American Chemical Society contact lens safety considerations
http://pubs.acs.org/hotartcl/chas/97/mayjun/con.html
- American Chemical Society
Less Is Better
http://membership.acs.org/c/ccs/pubs/less_is_better.pdf
- American Chemical Society
Read the Label
http://membership.acs.org/c/ccs/pub_7.htm
- American Chemical Society
Safety Audit Inspection Manual
http://membership.acs.org/c/ccs/pub_4.htm
- American Chemical Society
Guide for Chemical Spill Response Planning in Laboratories
http://membership.acs.org/c/ccs/pubs/spill_guide.htm
- American Red Cross
www.redcross.org
- ANSI American National Standards Institute
www.ansi.org
- Centers for Disease Control and Prevention
www.cdc.gov
- Contact Lenses in the School Laboratory Safety Discussion
www.flinnsci.com/Sections/Safety/eyeSafety/contactLenses.asp
- Directory of States with OSHA plans
www.osha.gov/fso/osp/index.html
- International Code Council—BOCA Standards
http://aec.ihs.com/abstracts/boca-standards.jsp
- National Fire Protection Association
www.nfpa.org
- National Institute for Occupational Safety and Health
www.cdc.gov/niosh/homepage.html
- National Science Teachers Association
Minimum Safety Guidelines for Presenters and Workshop Leaders
http://ecommerce.nsta.org/sessions/pdf/440.pdf

Live Long and Prosper

And Remember You Are Responsible

With each passing year, the role of the introductory college science instructor becomes more important to society as the decisions individual members must make each day increasingly require an understanding of how to collect, evaluate, and interpret data. Many of your students will use your course as the springboard to careers requiring knowledge of basic scientific skills and concepts. But even for students who have no interest in a career in science, the skills and experience they acquire in your laboratory investigations can inform their decision-making ability as consumers, voters, and leaders. With more adults skilled in the scientific way of knowing, we'll all live long and prosper.

Four Ps for Professionals

Research studies have firmly established that for students to reason scientifically and truly master the skills and concepts of science, they must directly engage in laboratory investigations. Whether you conduct the labs that correspond to your lecture sections personally or share the responsibility with colleagues, you have both professional and legal responsibility to ensure that the activities are appropriate and safe. You can assume this responsibility because you:

- *Prepare:* Your formal education and studies are just the beginning of your professional training and development. You constantly keep up-to-date with education by participating in professional organizations and by reading journals and research reports not only in your subject area but also in educational research, instructional strategies, and laboratory safety. You make yourself familiar with licensing requirements and regulations governing careers related to the courses you teach. You read and incorporate all institutional policies and procedures that relate to your duties and responsibilities, reviewing and filing new information and bulletins that update manuals and other official documents.

- *Plan:* You take the time to consider what you are required to teach and the best strategies to ensure your students can learn effectively and safely. You provide easy, confidential means for students to inform you of their special needs or constraints. You provide students and your assistants with written information on the course and the labs ahead of time so that they can review procedures and prepare properly. You review your own plans and syllabus at the end of each term and make appropriate modifications to keep your course and labs up to date.
- *Prevent:* You take the time to assess hazards, review procedures for accident prevention, and practice emergency response procedures. You read journals and check the material safety data sheets (MSDSs) on the chemicals you buy. For hazards that might be anticipated, you teach and review safety procedures with every student and assistant. You post safety signs and keep copies of safety information. If you detect safety hazards you cannot ameliorate, you put your concerns and requests for assistance or changes in writing to the appropriate administrator and modify your plans until the situations are remedied.
- *Protect:* You check your facilities for the presence and accessibility of correctly operating safety equipment and protective devices. You count the protective devices you have, such as eye protection and aprons, to ensure you have enough in good condition for everyone who needs them. You demonstrate and instruct students and assistants on the proper use of safety equipment and protective gear. You keep dated, easily understood records of safety lessons and instructions to ensure no one has missed getting the information. You insist that everyone in the room, including yourself and all visitors, use protective devices.

Broadening Your Definition of Safety

The previous chapters of this book are not intended to be an exhaustive manual of everything you must do or know to ensure safe science investigations. Rather, they are meant to sharpen your observational skills so you recognize the issues and circumstances that require your attention and planning in order for you to conduct science inquiry safely. As a science instructor, you need to consider safety as broadly as possible. No single book or series of books can anticipate all the safety issues that can arise in an active science program. Nor can any book, this one included, anticipate what new information and technology will render the advice given inaccurate or incorrect. It is the habit of observing and thinking about science lessons with common sense and safety in mind that will keep you and your students safe.

A Diversity of Needs

College science courses are not just for those interested in pursuing careers in the sciences and related vocations. It is important to offer all students a full range of investigative science experiences that lead to inquisitive habits of mind and safe practices in everyday life. Twenty-first century college graduates will be faced, almost daily, with questions and decisions best made when they are informed by the ability to observe, gather, and analyze data, draw conclusions based on empirical evidence, and critically challenge accepted theories. Science instruction and laboratory experience for all students might well be considered a safety requirement for a democratic society.

Whether your students have formally requested accommodation, or simply manifest diverse learning styles, it's important to always consider how to appeal to all of their strengths. Reconsider the suggestions in Chapter 2. They aren't just for disabled students but also can help you create a program that will enable every student to achieve a personal best.

Guest Instructors, Laboratory and Student Assistants

The actions taken by guest instructors and assistants to your program are part of your responsibility. These people are considered to be implementing your plans and your instructions. Student assistants are seldom qualified to conduct laboratories without your supervision. Never leave a class or laboratory with only a student assistant in charge. If a guest instructor plans a demonstration, you should review all plans ahead of time and ensure that the work will conform to your own safety standards.

Your Classroom—Your Castle

Friends, guests, and other observers may be welcome at your lectures and discussions but are best excluded from laboratory sessions. They are not likely to have had all the careful and documented safety instruction you have provided to your students and cannot be expected to recognize the hazards or safety procedures you have highlighted previously.

Discourage students from even asking to bring "my cousin from Peoria ..." into the laboratory. Laboratory work usually generates more excitement and more physical movement than most other activities, so it's best to avoid adding another body to the space, or another distraction to the group.

Your institution may require that your laboratory work be observed for performance evaluation or as part of some other institutional priority. Because you are responsible for the safety of everyone present during laboratory investigations, you have the right to insist everyone acknowledge and comply with all safety procedures prior to joining or observing your class.

Legal Responsibilities

Some instructors who are competent in laboratory work and instruction have drawn back from really interesting experiences because of fear of litigation. Although the fear may be real, it is insufficient reason to deprive students of the critical laboratory work they need to complete their science education.

In today's litigious society, you cannot prevent a lawsuit from being filed. Our legal system provides the opportunity for people to take their grievances to court. There are, however, many things you can do to avoid behaving negligently or being found at fault or liable. Your best defense is to have thorough written records showing that you have prepared and conducted your work in a prudent and reasonable manner, that you have taken steps to avoid unnecessary risks, and that you have diligently instructed students on the safe conduct of their investigations.

The Jargon

If an accident occurs, an instructor can be engulfed in a maze of legal terms and procedures. You'll want legal support immediately. Understanding the basics of civil law can help you avoid its pitfalls. Here are some simplified definitions:

- *Misfeasance:* Performance of a lawful action in an illegal or improper manner. In a science activity, this might result from unintentionally using an incorrect chemical for an experiment or too much of the correct chemical. It might include selecting an activity inappropriate for the students to whom it was assigned or providing incorrect instructions. The further you deviate from your department's and your own written course objectives, the greater the risk you take on.
- *Negligence:* Failure to exercise appropriate care. (This could be called nonfeasance.) This could include failure to warn students of safety hazards, failure to provide eye protection or fire equipment, or failure to practice a standard fire drill exit procedure. If you had eye protection available but did not make sure your students were using the protection when needed, this could be considered an omission and result in liability. Allowing laboratory work to be conducted without the presence of a properly trained supervisor might also lead to an accusation of negligence.
- *Malfeasance:* Intentional wrongdoing, deliberate violation of a law or standard, or mismanagement of responsibilities. Ignorance is no excuse. If there is a governing law or regulation, or local written policy, you are responsible for knowing about it and conforming. Regulations can have the effect of law even if there is no penalty statute. In the event of an accident that might have been prevented with the required safety procedure, your failure to instruct students and assistants in the procedure may not result in a criminal charge against you, but it could be construed as malfeasance for which you could be found liable.
- *Tort:* A wrong you do to someone. If you give students the wrong instructions or fail to provide appropriate safe instructions for performing an activity and the activity results in accident or injury, your action is a tort.
- *Liability:* The potential responsibility for damage; the civil obligation not to commit a wrong.

The Best Defense

Many states limit litigation against K–12 instructors to cases of gross negligence. That practice is seldom extended to college instructors. Even if it were in your state, the cost and complexity of determining what constitutes gross negligence isn't something you'd want to experience. In addition, anything you do outside the official scope of your employment (consulting, counseling students, volunteering for a high school presentation) probably is not covered by your institution's liability policy.

You can reduce your liability dramatically by adopting some relatively straightforward practices. Here are some tips:

- Document your preparation for safety. Subscribe to journals, read books, and take workshops or seminars to keep up-to-date. Online courses on safety may be convenient and timely.
- As part of your lesson plans and records, document safety instruction. Make sure you keep records of follow-up with safety lessons for students who are absent. Repeat the lessons individually with students who transfer into the class. Keep dated, easily read records of rules given to students and corrective action with rules violators.
- Do not leave the premises when you are responsible for students—*ever*—not even if you have a laboratory assistant. Do not permit students to work where they are out of your sight and supervision or send them on errands with hazardous materials or equipment. This can result in a serious liability.
- Put your safety needs in writing. Don't just complain: Explain why equipment and maintenance needs are necessary. Buy your department chair or dean a copy of this book, if necessary. Associate each request with the laboratory exercises that would be possible if it were fulfilled. Follow up on your requests.
- Do not engage in any activity involving a reported safety hazard until the hazard is mitigated. Modify your labs until the corrections have been made and keep a record of the changes you are making. That doesn't mean do nothing. Use smaller quantities. Break labs into subunits. Do part one week, and another part the next. Substitute computer-based video clips or do a demonstration.
- If you have an accident, collect your records and document your recollections immediately. Make sure you have the advice, and possibly the presence, of an attorney who represents your interests before making any statements to anyone else. Consult your faculty union or association as well as the administration. If there is any chance that your interests are not the same as your institution's interests, you should also consider retaining the services of your own attorney.

> **Documents to Keep Up-to-Date**
>
> - Institutional policy manuals and all subsequent policy communications
> - Course descriptions and syllabi
> - Plan book or calendar for lectures and laboratories
> - Attendance and grade books—with correlations to the plan book and records of safety lessons for each individual student
> - Inventory of materials and equipment
> - Maintenance requests
> - Purchase requisitions with MSDSs
> - Records and notes of professional development activities

Insurance

Lawsuits are expensive even if they are dismissed or you are found not to be responsible. It can take months and cost tens of thousands of dollars just to get a suit in front of a judge so you can have it dismissed. There will be accusations, investigations, and depositions. That's why insurance is vital.

Begin by thoroughly investigating what coverage you already have. Your college may have errors and omissions coverage for the institution and its employees. This can cover mistakes. If your institution is state supported, it may have some sovereign immunity, but that protection may not apply to you. Find out what the protections and exceptions are.

You may have some liability insurance through your union or professional association. In general, these policies will cover the preliminary costs of a lawsuit. They will get you a lawyer quickly, separate from the one that represents the institution, and get you preliminary advice. But these policies may exclude punitive damages, so the cost of your lawyer may be covered but not the biggest part of a future settlement.

Insurance policies may exclude coverage for any action that violates the law, regulations, or written policy. For example, if there is a regulation or institutional policy that requires you to conduct a fire drill within the first two weeks of each new class, and you fail to do so, you may find yourself without insurance coverage if there is a fire in the lab later in the term. If your state fire marshal has banned the use of alcohol burners in student labs and you have a lab accident involving the use of an alcohol burner, you may find yourself without insurance coverage.

Some institutional policies may cover employees, but not consultants, volunteers, or visitors. You might be covered as a college instructor, but not for your moonlighting as a curriculum consultant or volunteer work as a scout leader.

Be certain of your coverage by reading the pertinent policy and contract documents yourself or asking a professional to assist you.

Many instructors carry extra professional coverage through their professional association, homeowner's, or renter's policy. An umbrella liability policy that would provide representation for you if the worst occurs isn't as expensive as you might think and is essential for an instructor's peace of mind. You may be able to obtain an umbrella policy to extend coverage and back up other liability policies you may have. Your coverage might be a rider for professional liability on an existing policy, usually at very little additional cost. Extra coverage is highly recommended.

Take Heart

With all this talk of litigation and liability you may be asking, "Is the laboratory work worth it?" The answer is, "Yes, absolutely." An instructor who stays current, plans carefully, embeds safety training all term long, and properly supervises each student has little to fear. When you take time to prepare for conducting an active laboratory science program more safely, you not only assist your students in answering the questions on their next test, but you also prepare your students to answer future questions we have yet to imagine. Your habit of conducting work thoughtfully and more safely becomes the model for the conduct of your students at home and in their future endeavors. As the late teacher-astronaut Christa McAuliffe so elegantly put it: "I touch the future. I teach."

THE SAVVY SCIENCE INSTRUCTOR

Mr. H was rarely afraid. He climbed rock walls, canoed white water, and rode roller coasters. But the day he realized that part of his chemical inventory was missing, a frisson of fear ran through him.

He first suspected something was wrong when he prepped for a lab. A dustless circle in the corrosives cabinet told him to check his inventory. A few keystrokes and his inventory program told him what had happened: A liter of concentrated hydrochloric acid had disappeared.

Mr. H immediately notified the department chair and the facilities manager. He pulled the MSDS on the missing chemical from the computer and verified that the stock was fresh and concentrated—hazardous in the wrong hands. Together they checked the locks. There was no sign of forced entry. But the acid was most assuredly gone.

Campus security was called in to investigate. "Who might have had keys to this storeroom?" she asked. "Only science instructors," both Mr. H and his department chair answered simultaneously. Almost immediately, they also realized that former science instructors might still have keys. They made a quick list. Dr. J had retired, and turned in his keys right before leaving. But how about

(cont. on next page)

11

(cont. from preceding page)

Ms. K? She had once taught natural science laboratory, but was now teaching in her major, art.

A flash of insight prompted the department chair to make a phone call. Yes, Ms. K had entered the storeroom and taken the acid to etch jewelry for a project she was creating for a charity benefit. The incident prompted a change in procedure. From now on, all keys would be returned and accounted for at the end of each school year. Thanks to good records and alert monitoring, a mystery was solved without risk to staff or students.

Connections

- Classroom Science and the Law
  Laboratory Safety Institute
  www.labsafety.org
- Law Dictionary
  http://dictionary.law.com/
- Overcrowding in the laboratory
  www.flinnsci.com
- Sample of Chemical Classroom Accidents
  www.safechem.com/Campusafe/accident.htm

Conclusion

Review the Basics

Teaching is a multifaceted, complex process. You make dozens of instructional decisions each hour. You manage an environment, a curriculum, and a community of learners. Safety issues must be an integral part of science instruction and embedded in every laboratory and field activity. This book has taken a broad view of factors that should be considered when preparing for a safe investigative science program. To relate these ideas to reality, we have included anecdotes—some almost verbatim and some composites—of actual experiences of instructors and supervisors. They are meant to illustrate both the risks and the opportunities in the familiar environment of your science room.

We hope these ideas have sharpened your perceptions, raised your safety antennae, and also demonstrated that the best safety advice is heightening your awareness rather than merely following rules that may become outdated. Here are some general principles that may be drawn from each of the chapters:

1 Setting the Scene

- Rather than confining safety issues to an introductory unit, introduce safe procedures with each exploration and repeat relevant safety instructions with each subsequent activity.
- Model the safe behaviors you expect your students to practice.
- Have students share responsibility for monitoring safe procedures so that safe work habits become second nature.
- Make administrators and facilities staff part of your safety team. Educate them about the conditions and facilities needed to teach science safely.
- Modify your curriculum to conform to the conditions of your facilities and the nature of your students and classes.

2 Communities of Learners

- In planning laboratory activities, provide modifications and accommodations to ensure they are maximally accessible to all students.
- Eliminate barriers in your room, including removing clutter and freeing up space.

- Communicate with administrators and special-needs staff to obtain support and suggestions for modifying your program for special-needs and disabled students.
- Minimize the potential for inattention with shorter and more specific tasks.
- Request appropriate support staff and aides for students who need them. Plan time to train these assistants.
- Investigate the use of adaptive equipment and technologies with consultants and district support staff.
- Read and incorporate individual education plan (IEP) recommendations for dual-enrollment students who have IEPs.

3

Where Science Happens

- Conduct science activities in facilities that provide adequate space and ventilation for safety.
- Check for appropriate utilities and safety equipment for all laboratory activities. If items or conditions are not adequate, modify your curriculum to ensure that the only activities you conduct are activities that can be done safely.
- Select furniture that is stable but easily rearranged, and provide unobstructed flat work surfaces.
- Ensure that safety equipment such as fire extinguishers, fire blankets, safety showers, and eyewashes are operational and accessible.
- Check that electrical service and wiring to your room are well maintained, provide adequate amperage, and include appropriate ground-fault circuit interrupter (GFCI) protection.

4

Finders Keepers

- Use well-organized and clearly marked open shelves to store supplies and equipment that students are to obtain and return on their own. Arrange the material so that a missing item is easily spotted without counting individual items.
- Include enough locked storage, inaccessible to students, for valuable and fragile supplies and equipment as well as materials too hazardous for direct student access.
- Store all chemical stocks in locked cabinets and locked chemical storerooms appropriately equipped and off-limits to students.

- Keep incompatible chemicals separated—arrange storage by chemical properties, not by alphabetical order.
- Maintain accurate, up-to-date, and regularly reviewed chemical inventories.
- Maintain materials safety data sheets (MSDSs) for every product—one set in the office and one set at the storage-and-use location.
- Prepare materials and supplies for science activities in an adequate, ventilated, and well-lit preparation space away from students and other traffic.
- Clear out excess supplies, equipment, and furniture regularly as a vital part of safe practice.

5 Lively Science

- Maintain living cultures to provide students with opportunities for activities involving observation and care of living organisms and biological systems.
- Choose organisms appropriate for the space and time available. Ensure that adequate safety and security are available and that you can maintain the organism in a healthy environment over weekends and extended vacation periods.
- Begin with simple organisms—plants and invertebrates—before trying to maintain more complex and difficult ones.
- Do not bring wild or feral animals, dead or alive, into the classroom.
- Avoid organisms that are toxic, highly allergenic, or temperamental.
- Provide information about your cultures to students and colleagues who share facilities. Be aware of unusual allergies.
- Before introducing a living organism in shared spaces, ensure that all users agree to provide appropriate safety and security.
- Do not release, introduce, or plant non-native species in the open environment. Avoid the use of exotic species if at all possible.

6 Modern Alchemy

- Emphasize careful scientific process skills over drama.
- Use microscale experiments for safety and to encourage careful observation.
- Choose less toxic and less hazardous options over traditional labs now known to be dangerous.

- Maintain a minimal quantity and variety of chemicals—less is better.
- Require the use of appropriate safety equipment by all persons—students and adults—at all times.
- Consider the problems and costs of disposal before purchasing any chemical reagent.
- Use professional hazardous waste removal services as appropriate.

7 Striking Gold

- Before beginning activities, instruct students on safe procedures and safety mechanisms for tools and equipment.
- Never allow the tasting of specimens.
- Avoid contact with and use of contaminated soils, and insist on proper hand washing.
- Never permit direct observation of the Sun.
- Review rules and safety procedures prior to field studies.

8 Falling for Science

- Before beginning activities, instruct students on safe procedures and safety mechanisms for tools and equipment.
- Use light and sound experiments as an opportunity to discuss preventing damage to eyes and ears.
- Make sure all electrical connections are safe and conform to code.
- Plan ahead so your room won't have physical barriers such as loose cords and other tripping and falling hazards.

9 The Great Outdoors

- Link field trips and field studies to curriculum goals.
- Preview the site and abutting properties before planning your field study.
- Determine proper clothing and footwear for the site and activities planned.
- Meet with cooperating resource people to plan activities.

- Orient and train all chaperones in your planned activities and in safety precautions.
- Plan appropriate accommodations for special-needs and physically disabled students.
- Encourage student participation and responsibility in planning field studies.

10 The Kitchen Sink

- Minimize the use of carpets and upholstered furniture to reduce the growth and harboring of dust mites, mold spores, and other allergens.
- Regularly and carefully clean living cultures.
- Check for the presence of heavy-metal contamination from prior activities—remove and appropriately dispose of all mercury and mercury-based instruments.
- Review appropriate clothing, covering, and protective eyewear for laboratory work.
- Preview and vet internet resources.
- Protect confidential student information and obtain consent forms before photographing students.
- Never prepare food or eat in science lab areas.
- Participate in first aid training and crisis response training.
- Provide convenient and confidential means for students to communicate with you.
- Plan with others for appropriate class sizes and scheduling, and adequate space and materials for an investigative science program.
- Make safety and security a significant concern of all staff.

11 Live Long and Prosper

- Prepare, plan, prevent, and protect.
- Become familiar with institutional policies and federal, state, and local laws.
- Document all safety-related activities including instructions provided to each student and assistant.
- Assess your safety lessons, and keep detailed records.
- Take responsibility to supervise and train your assistants and volunteers.
- Make sure you are adequately protected with liability insurance.

References

Biehle, J., L. Motz, and S. West. 1999. *NSTA guide to school science facilities.* Arlington, VA: NSTA Press.

Fathman, A. K., and D. T. Crowther, eds. 2006. *Science for English language learners: K–12 classroom strategies.* Arlington, VA: NSTA Press.

McCullough, J., and R. McCullough. 2000. *The role of toys in teaching physics.* College Park, MD: American Association of Physics Instructors.

National Research Council (NRC). 1996. *National Science Education Standards.* Washington, DC: National Academy Press. Online version at *www.nap.edu/books/0309053269/html/index.html.*

Russell, H. R. 2001. *Ten-minute field trips.* Arlington, VA: NSTA Press.

Sarquis, M. 2000. *Building student safety habits for the workplace.* Middletown, OH: Terrific Science Press.

Texley, J., and A. Wild, eds. 2000. Appendix C in *NSTA pathways to the science standards—high school edition.* Arlington, VA: NSTA Press.

Waugh, M. 2003. The plague generation today. *The Science Instructor* 70 (7): 42–45.

Web Resources

As teachers of science we promote a healthy dose of skepticism. We encourage our students not just to seek information but also to think for themselves and question what is presented. We encourage you the reader to use that same caution in using the websites we have listed. To our knowledge they are as they profess to be; however, we urge you to make your own decisions about their validity and usefulness. Websites change with time. We offer these sites not as a bona fide list, one of certainty, but as a list of starting points—possible sources of information.

The websites accessed through NSTA's SciLinks program are continually reviewed and updated.

Chapter 1: Setting the Scene

American Association for the Advancement of Science
www.aaas.org

American Chemical Society and ACS Board–Council Committee on Chemical Safety. 2001. Chemical safety for instructors and their supervisors. Washington, DC: ACS. Available in PDF format at
http://membership.acs.org/c/ccs/pubs/Chemical_Safety_Manual.pdf

Flinn Scientific Catalog/Reference Manual

Flinn Scientific, Inc. , 2002. Batavia, IL
www.flinnsci.com

Laboratory Safety Institute, James A. Kaufman, President
www.labsafety.org/about.htm

MSDS for Infectious Substances, Health Canada
www.hc-sc.gc.ca/pphb-dgspsp/msds-ftss/index.html

National Association of Biology Teachers
www.nabt.org

National Board for Professional Teaching Standards
www.nbpts.org

National Research Council (NRC). 1996. *National Science Education Standards.* Washington, DC: National Academy Press.
www.nap.edu/openbook/0309053269/html

National Science Teachers Association
www.nsta.org

Vermont Safety Information Resources, Inc. (source for chemical MSDS).
www.siri.org
www.hazard.com

Chapter 2: Communities of Learners

ADA, The Americans With Disabilities Act
www.usdoj.gov/crt/ada/adahom1.htm
www.usdoj.gov/crt/ada/pubs/ada.txt

American Chemical Society. 2001, 4th ed. Teaching chemistry to students with disabilities. Washington, DC: ACS.
http://membership.acs.org/C/cwd/teachchem4.pdf

Barrier Free Education
http://barrier-free.arch.gatech.edu

CEC, The Council for Exceptional Children
www.cec.sped.org

IDEA, Individuals With Disabilities Education Act
www.ed.gov/offices/OSERS/Policy/IDEA/index.html

National Research Council (NRC). 1996. *National Science Education Standards*. Washington, DC: National Academy Press.
www.nap.edu/openbook/0309053269/html

West Virginia University, Inclusion in Science Education for Students with Disabilities
www.as.wvu.edu/~scidis

U.S. Census. Language Use and English-speaking Ability: 2000
www.census.gov/prod/2003pubs/c2kbr-29.pdf

National Clearinghouse for English Language Acquisition
www.ncela.gwu.edu

National Science Teachers Association. 2000. Position Statement on Multicultural Education
www.nsta.org/positionstatement&psid=21

SCILINKS THE WORLD'S A CLICK AWAY
Topic: learners with disabilities
Go to: *www.scilinks.org*
Code: SSCC16

Chapter 3: Where Science Happens

SCI LINKS. THE WORLD'S A CLICK AWAY
Topic: fire extinguishers
Go to: *www.scilinks.org*
Code: SSCC44

American Society for Microbiology hand washing information
www.microbe.org/washup/wash_up.asp

ANSI Z87.1, American National Standards Institute
www.ansi.org
www.coltslaboratories.com/

Environmental Protection Agency IAQ INFO 800-438-4318
www.epa.gov

FDNY New York City, New York Fire Department
www.nyc.gov/html/fdny/html/safety/extinguisher/classes.shtml

Fume Hoods Fact Sheet, Office of Environmental Health and Safety, University of California Berkeley, 11 September 2003
www.ehs.berkeley.edu/pubs/factsheets/09fumehd.html

MSDS (materials safety data sheet)
www.flinnsci.com
www.esf.uvm.edu/uvmsafety/labsafety/chemsafety/netmsds.html
http://msds.ehs.cornell.edu/msdssrch.asp

National Fire Protection Association
www.nfpa.org/index.asp

National Institute for Occupational Safety, Chemical Safety
www.cdc.gov/niosh/topics/chemical-safety

NFPA 45, Standard on Fire Protection for Laboratories Using Chemicals
www.nfpa.org/Codes/NFPA_ Codes_and_Standards/List_of_NFPA_documents/NFPA_ 45.asp

NFPA 101, Life Safety Code
www.nfpa.org/BuildingCode/AboutC3/NFPA101/nfpa101.asp

NFPA 5000, Building Construction and Safety Code
www.nfpa.org/catalog/product.asp?pid=500003&src=nfpa&link_type=buy_box&order_ src=A292

Occupational Safety and Health Administration
www.osha.gov

OSHA, Hazard Communication
www.osha.gov/SLTC/hazardcommunications/index.html

OSHA Regulations (Standards—29 CFR) "Laboratory Standards"
Occupational exposure to hazardous chemicals in laboratories. 1910.1450.

www.osha.gov/pls/oshaweb/owadisp.show_document?p_table=STANDARDS&p_id=10106

OSHA Regulations (Standards—29 CFR) "Chemical Hygiene Officer and Program" National Research Council Recommendations Concerning Chemical Hygiene in Laboratories (Non-Mandatory). 1910.1450 App A.
www.osha.gov/pls/oshaweb/owadisp.show_document?p_table=STANDARDS&p_id=10107

SIRI, Vermont Safety Information Resources, Inc.
www.siri.org

Chapter 4: Finders Keepers

American Chemical Society, "Less is Better"
http://membership.acs.org/c/ccs/pubs/less_is_better.pdf

Cornell University Environmental Health
www.ehs.cornell.edu

U.S. Environmental Protection Agency is responsible for hazardous waste
www.epa.gov/ebtpages/wastes.html

SIRI, Vermont Safety Information Resources, Inc.
www.siri.org

Flinn Chemical and Biological Catalog Reference Manual
www.siri.org

SCILINKS. THE WORLD'S A CLICK AWAY
Topic: chemical safety
Go to: *www.scilinks.org*
Code: SSCC50

SCILINKS. THE WORLD'S A CLICK AWAY
Topic: hazardous waste
Go to: *www.scilinks.org*
Code: SSCC61

Chapter 5: Lively Science

SCILINKS. THE WORLD'S A CLICK AWAY
Topic: bacteria
Go to: *www.scilinks.org*
Code: SSCC75A

SCILINKS. THE WORLD'S A CLICK AWAY
Topic: protista
Go to: *www.scilinks.org*
Code: SSCC75B

SCILINKS. THE WORLD'S A CLICK AWAY
Topic: Fungi
Go to: *www.scilinks.org*
Code: SSCC76

SCILINKS. THE WORLD'S A CLICK AWAY
Topic: invertebrates
Go to: *www.scilinks.org*
Code: SSCC77

Topic: viruses
Go to: *www.scilinks.org*
Code: SSCC80

Topic: AIDS
Go to: *www.scilinks.org*
Code: SSCC81

Topic: salmonella
Go to: *www.scilinks.org*
Code: SSCC84

Topic: hazardous waste
Go to: *www.scilinks.org*
Code: SSCC85

American Society for Microbiology
www.microbe.org

Biosafety in the Laboratory: Prudent Practices for Handling and Disposal of Infectious Materials
http://books.nap.edu/catalog/1197.html

Biosafety in Microbiological and Biomedical Laboratories (BMBL) 4th Edition (n.b., This volume issued jointly by the CDC and NIH is being updated; search for the 5th edition)
www.cdc.gov/od/ohs/biosfty/bmbl4/bmbl4toc.htm

Centers for Disease Control and Prevention (CDC)
www.cdc.gov/healthypets/

Flinn Scientific Catalog/Reference Manual. 2002. Batavia, IL: Flinn Scientific, Inc.
www.flinnsci.com

Guidelines for Ethical Conduct in the Care and Use of Animals
www.apa.org/science/anguide.html

Hand washing. PMC ISLA Health System—Guam
www.pmcguam.com/resource/handwash.htm

Institute for Laboratory Animal Research
http://dels.nas.edu/ilar_n/ilarhome/reports.shtml

Institutional Animal Care and Use Committees (links to extensive information)
www.IACUC.org

MSDS for Infectious Substances, Health Canada
www.phac-aspc.gc.ca/msds-ftss/index.html

National Research Council Guide for the Care and Use of Laboratory Animals
http://books.nap.edu/catalog/11138.html

http://books.nap.edu/catalog/5140.html
Precollege animal care brochure
http://dels.nas.edu/ilar/ilarhome/Principles_Guidelines.pdf

PTC Tasting physchem
http://physchem.ox.ac.uk/MSDS/PH/1-phenyl-2-thiourea.html

Chapter 6: Modern Alchemy

American Chemical Society and ACS Board–Council Committee on Chemical Safety. 2001. Chemical safety for instructors and their supervisors.
http://membership.acs.org/c/ccs/pubs/Chemical_Safety_Manual.pdf

Cornell University
http://msds.ehs.cornell.edu/msdssrch.asp

Fisher Scientific
www1.fishersci.com/index.jsp

Microscale Gas Chemistry. Ideas for Microscale Chemistry
http://mattson.creighton.edu/LabExperiments.html
http://mattson.creighton.edu/HowToPrepGases/Index.html

National Microscale Chemistry Center
http://microscale.org/

Samples of Chemistry Accidents. SafeChem.
www.safechem.com/Campusafe/accident.htm

SIRI, Vermont Safety Information Resources, Inc.
www.hazard.com

Chapter 7: Striking Gold

The Microbial World
http://helios.bto.ed.ac.uk/bto/microbes/winograd.htm

National Aeronautics and Space Administration
www.nasa.gov

National Oceanic and Atmospheric Administration
www.noaa.gov

Standard 29CFR1910.95—Occupational Noise Exposure

SCiLINKS THE WORLD'S A CLICK AWAY
Topic: astronomy
Go to: *www.scilinks.org*
Code: SSCC112

Occupational Safety and Health Administration (OSHA), U.S. Department of Labor
www.osha.gov

U.S. Geological Survey
www.usgs.gov

Chapter 8: Falling for Science

SCILINKS THE WORLD'S A CLICK AWAY

Topic: noise pollution
Go to: *www.scilinks.org*
Code: SSCC118

SCILINKS THE WORLD'S A CLICK AWAY

Topic: electricity
Go to: *www.scilinks.org*
Code: SSCC120A

SCILINKS THE WORLD'S A CLICK AWAY

Topic: electrical safety
Go to: *www.scilinks.org*
Code: SSCC120B

American Association of Physics Instructors. Safety in Physics Education. 2002.
www.aapt.org/Store/products.cfm?ShowAll=1

Ground Fault Interrupter Circuit (GFCI)
www.electrical-contractor.net/ESF/GFCI_Fact_Sheet.htm

The International Electrical Safety Foundation
www.esfi.org/index.php.

Physics Safety Issues. The Catalyst.
www.thecatalyst.org/hwrp/safetymanual/index.html

U.S. Department of Labor, OSHA—Laser Hazards, standards
www.osha.gov/SLTC/laserhazards/index.html

Chapter 9: The Great Outdoors

Amusement park physics. 1997–2003. Annnenberg/CPB
www.learner.org/exhibits/parkphysics/

Centers for Disease Control and Prevention
www.cdc.gov/travel/yb/index.htm

Department of State
www.state.gov/travel/

Kids' Lightning Information and Safety
www.kidslightning.info/zaphome.htm

SCILINKS THE WORLD'S A CLICK AWAY

Topic: allergies
Go to: *www.scilinks.org*
Code: SSCC140

Robertson, W.C. 2001. *Community connections for science education: Building successful partnerships.* Arlington, VA: NSTA Press
http://store.nsta.org/showItem.asp?product=PB160X1&session=4183699CAA084FEC99AB1AF464AFF492

Topic: preserving ecosystems
Go to: *www.scilinks.org*
Code: SSCC148

Chapter 10: The Kitchen Sink

American Chemical Society and ACS Board–Council Committee on Chemical Safety. 2001. Chemical safety for instructors and their supervisors. Washington, DC: ACS.
http://membership.acs.org/c/ccs/pubs/Chemical_Safety_Manual.pdf

American Chemical Society
Contact lens safety considerations
http://pubs.acs.org/hotartcl/chas/97/mayjun/con.html

American Chemical Society
Less Is Better
http://membership.acs.org/c/ccs/pubs/less_is_better.pdf

American Chemical Society
Read the Label
http://membership.acs.org/c/ccs/pub_7.htm

American Chemical Society
Safety Audit Inspection Manual
http://membership.acs.org/c/ccs/pub_4.htm

American Chemical Society
Guide for Chemical Spill Response Planning in Laboratories
http://membership.acs.org/c/ccs/pubs/spill_guide.htm

American Red Cross
www.redcross.org

ANSI American National Standards Institute
www.ansi.org

Centers for Disease Control and Prevention
www.cdc.gov

Contact Lenses in the School Laboratory Safety Discussion
www.flinnsci.com/Sections/Safety/eyeSafety/contactLenses.asp

Directory of States with OSHA plans
www.osha.gov/fso/osp/index.html

International Code Council—BOCA Standards
http://aec.ihs.com/abstracts/boca-standards.jsp

National Fire Protection Association
www.nfpa.org

National Institute for Occupational Safety and Health
www.cdc.gov/niosh/homepage.html

National Science Teachers Association
Minimum Safety Guidelines for Presenters and Workshop Leaders
http://ecommerce.nsta.org/sessions/pdf/440.pdf

Occupational Safety and Health Administration
www.osha.gov

OSHA standard concerning food and drink in laboratories
www.osha.gov/pls/oshaweb/owadisp.show_document?p_table=STANDARDS&p_id=9 79029CFR1910.141g(2&4)

Chapter 11: Live Long and Prosper

Classroom Science and the Law
Laboratory Safety Institute
www.labsafety.org

Law Dictionary
http://dictionary.law.com

Overcrowding in the laboratory
www.flinnsci.com

Sample of Chemical Classroom Accidents
www.safechem.com/Campusafe/accident.htm

Glossary

Page numbers for discussion of the topic are indicated in parentheses. Use the index to find specific text using the terms.

ADA: The Americans With Disabilities Act of 1990. This act requires that physical and other accommodations be made to enable all individuals to have access to physical facilities, education, and employment. (Chapter 2, p. 18)

ANSI: The American National Standards Institute is a private, nonprofit organization that administers and coordinates the U.S. voluntary standardization and conformity assessment system.

ANSI Z186.1: Laser safety standards accredited by the American National Standards Institute. A revision of the 1993 document was due to be released in early 2006.

ANSI Z358.1: Emergency shower and eyewash standards accredited by American National Standards Institute.

ANSI Z87.1-2003: Officially called the "American National Standards Practice for Occupational and Educational Personal Eye Protective Devices," this is a voluntary standard for impact-resistant safety eyewear. The Occupational Safety and Health Administration (OSHA) has made the standard a requirement. (Chapter 10, p. 161)

bloodborne disease: Human illnesses that can be transmitted via contact with infected human blood. These include hepatitis B, hepatitis C, and HIV/AIDS. (Chapter 10, pp. 156–157)

BOCA: Building Officials and Code Administrators International. The organization is a founding member of the International Code Council (ICC). Its guidance is invaluable in the design and construction of science facilities.

carcinogen: A substance that can cause cancer. Many laboratory chemicals, such as formaldehyde, can be carcinogenic and should be avoided or used very carefully. Material safety data sheets (MSDSs) and the Merck Index can be used to check for carcinogenic properties of chemicals.

CDC: Centers for Disease Control and Prevention. A federal agency responsible for tracking, investigating, and recommending methods to contain the incidence of diseases in the population, including diseases that may be acquired outside the United States. (Chapter 9, p. 134)

combustible: A substance that ignites and burns easily, such as paper or cardboard.

course platform: Sofware that permits teacher and student interaction via the internet. Written and graphic learning materials can be placed online by instructor and accessed by students. Examples are Blackboard or WebCT.

conjunctivitis: Highly infectious inflammation of the conjunctiva of the eye that can be caused by either bacteria or viruses. Commonly referred to as *pink eye,* it can be easily transmitted between users of safety eyewear or optical eyepieces. (Chapter 10, p. 162)

CORI check: Criminal offender record information check. Many states or institutions have regulations that require schools to perform a check of CORI databases before people are permitted to interact with students. This can include potential employees, adult volunteers, and guests.

cyanoacrylic glue: A class of glues known by a variety of trade names such as SuperGlue or Krazy Glue.

DEET: Diethyl toluamide. The active ingredient in many insect repellents. Some people may be allergic to the compound.

discrepant event: An unanticipated or unexpected result. Major scientific breakthroughs are often the result of scientists noticing a discrepant event that otherwise might be considered an accident or "failed" experiment. A classic example is Fleming's discovery of the anitibiotic effects of penicillin when the mold accidentally contaminated his culture dish.

endangered species: Organisms that are protected by federal, state, or local regulations because their populations are so low that survival of the species is in jeopardy. Protections include prohibitions on collection, killing, or disruption of breeding grounds or territories.

EPA: Environmental Protection Agency. There are both federal and state EPAs that promulgate and enforce regulations to protect the environment.

exotic species: Species that are not native to or naturally found in the environment they are in. These include imported plants and animals as well as plants and animals that may be from closer locations but introduced to environments that lack natural competitors or predators. (Chapter 5, pp. 79–80)

FERPA: Family Educational Rights and Privacy Act

filtering software: Computer programs that are installed to block access to internet websites that may be considered to have unacceptable content such as pornography, hate literature, or terrorist information. These programs usually operate by checking for words or phrases and are frequently ineffective or problematic because they do not block some sites that would be harmful or they block useful sites that legitimately use terms such as *vagina*. We strongly recommend that you closely supervise students' internet use and not rely on filtering software for protection. (Chapter 10, p. 165)

flammable: A substance that ignites easily and is capable of burning rapidly, such as alcohol or gasoline.

GFCI: Ground fault circuit interrupter. A device that detects and disrupts improper electric flows such as found in a "short circuit" when an electrical appliance is dropped into water. All electrical circuits in science areas should be GFCI-protected. (Chapter 8, p. 119)

hazardous waste: Chemical and biological waste that requires special handling—usually governed by federal, state, or local regulation—for disposal. This includes agents that can cause harm to humans, animals, plants, or the environment. (Chapter 4, p. 61)

hepatitis: Several infectious diseases of the liver, some of which are transmitted by human body fluids or food and water. In some cases, infectious carriers do not show active signs of the disease. Standard Precautions and thorough hand washing should be used to prevent transmission. (Chapter 10, p. 157)

HIV: Human immunodeficiency virus, a retrovirus that is the causative agent of AIDS (acquired immunodeficiency syndrome).

human body fluids: Blood, saliva, vomitus, semen, and other fluids produced by the human body. Most are known to transmit disease and must be handled with Standard Precautions. (Chapter 10, p. 156)

HVAC: Heating, ventilating, and air conditioning.

IDEA (PL 94-142): The Individuals With Disabilities Education Act, also known as Public Law 94-142. IDEA pertains to students who have not completed their high school education and mandates that those students with disabilities that prevent them from reaching their potential in education be provided with accessory materials and services. It also requires that, to the greatest extent possible, students with disabilities be educated with students who do not have disabilities. Community colleges may have to comply with IDEA if they have dual enrollees who have not completed their high school studies. (Chapter 2, p. 19)

impact-resistant goggles: Safety eyewear that provides impact protection. Impact-resistant goggles that conform to ANSI Z81.1 standards may or may not also provide chemical-splash protection. We recommend that all safety eyewear provide both chemical-splash and ANSI Z81.1-compliant impact protection. (Chapter 10, p. 161)

inflammable: Same as flammable, but often mistaken to mean not capable of burning. To avoid confusion, we suggest not using the term *inflammable*.

laboratory refrigerator: A refrigerator specifically designed not to emit sparks when cycling on so as to reduce explosion risks from volatile chemicals stored inside. It is significantly different from a household refrigerator. Any refrigerator—laboratory or household—used to store laboratory material should never be used to store food for human consumption.

laser: Light amplified by stimulated emission of radiation. Devices that emit this type of light must be used with extreme care to protect from exposure that can cause severe damage to eyes, skin, and other tissues. Refer to ANSI 186.1 standards for guidance. (Chapter 8, p. 126.)

LD_{50}: Lethal Dose 50. The dose of a toxic substance that results in the death of 50% of test animals. (Chapter 6, p. 99)

MSDS: Material safety data sheet. A standard document written in English and containing certain types of information available for every hazardous chemical manufactured or sold in the United States. It contains information that users and emergency personnel need to make appropriate decisions quickly. Federal and many state laws require that MSDSs be available all the time for every chemical you use or store. Administrative and noninstructional staff, teachers, students, and emergency responders must have easy access to them. (Chapter 4, p. 57)

mutagen: A substance or agent such as ultraviolet radiation that can increase the frequency of mutation in cells. The mutations may then result in cancer. Chemicals, such as some used in DNA studies, can be mutagenic and should be used with great care. Material safety data sheets (MSDSs) and the Merck Index can be used to check for mutagenic properties of chemicals.

National Science Education Standards: Goals for instruction, assessment, professional development, content, and programs in science education published in 1996 by the National Research Council, the administrative unit of the National Academy of Sciences. These standards were the result of a landmark effort by teachers, scientists, science educators, and other experts to define what science students should know and be able to do. They offer a coherent vision of what it means to be scientifically literate. (Chapter 2, p. 15)

NFPA: National Fire Protection Association. This international nonprofit group attempts to reduce fire and related hazards by gaining consensus and acceptance of codes and standards, research, training, and education.

NIOSH: National Institute for Occupational Safety and Health. Created by the Occupational Safety and Health Act of 1970, NIOSH is the federal agency established to research workplace safety and educate people about it. NIOSH is part of the Centers for Disease Control and Prevention in the Department of Health and Human Services.

nonlatex: Made of a rubberlike material other than latex rubber, often nitrile. Latex rubber has been found to be dangerously allergenic to some individuals and can result in reactions severe enough to cause death. For this reason, we recommend that all "rubber" items such as rubber gloves, rubber tubing, and rubber stoppers be nonlatex. Nonlatex gloves should be available for cleanup of human body fluids using Standard Precautions.

non-native species: Species that are not naturally found in the environment they are in. These include imported plants and animals as well as plants and animals that may be from closer locations but introduced to environments that lack natural competitors or predators. (Chapter 5, pp. 77, 79)

NRC: The National Research Council is the administrative unit of the National Academy of Sciences. It was the coordinating agency for the development of the National Science Education Standards. (Chapter 2, p. 28)

OSHA: Occupational Safety and Health Administration. Created by the Occupational Safety and Health Act of 1970, OSHA is the federal agency responsible for developing and enforcing workplace safety and health regulations.

PDA: Personal digital assistant. Blackberry is an example.

PDF: Portable document format. A software file format created by Adobe that is frequently used for electronic information exchange, because content can be transferred with integrity and used on a variety of platforms.

PFD: Personal flotation device. Life jackets are PDFs.

potable water: Water that is fit for human consumption. Many buildings with older plumbing have excessive concentrations of lead in the water, rendering it hazardous to drink or to use in food preparation.

prep room: A preparation room in which a science instructor can prepare and securely store materials for science laboratory activities, situated close to the science room and preferably with direct access to the science room via a separately lockable door. For safety and security reasons, chemical stock supplies should never be stored in preparation rooms. Students should not be allowed in preparation areas without the direct supervision of an instructor. (Chapter 3, p. 41, and Chapter 4, p. 50)

safety eyewear: This phrase is applies to a wide variety of equipment for protection of the eyes including safety glasses, safety goggles, "plant specs," and face shields. We recommend that safety goggles with both chemical-splash protection and ANSI Z87.1-compliant impact resistance be used in science labs. (Chapter 10, p. 161)

safety goggles: Safety eyewear that fits snugly to the face. These devices may provide chemical splash protection and/or impact protection. We recommend that safety goggles with both chemical-splash protection and ANSI Z87.1-compliant impact resistance be used in science labs. (Chapter 10, p. 161)

science room: In this book, *science room* refers to any room in which science classes are taught. In some cases, these are rooms outfitted for both lecture and laboratory work—the ideal. In other cases, a science room is for lecture or for laboratory work alone. A science room that is a general classroom used by science classes is a very risky practice. The term specifically excludes science storage rooms and science preparation rooms, because students should never be permitted in chemical storage rooms and, as a general rule, should not have access to science storage and preparation areas. (Chapter 3, p. 29)

SPF: Sun protective factor. A measure of the potency of sunscreens and sunblocks in blocking ultraviolet (UV) radiation. Sunscreens with SPFs between 15 and 30 will block most UV radiation, according to the National Cancer Institute. Some clothing may also be SPF rated. (Chapter 9, p. 141)

splash goggles: Safety eyewear that prevents chemical splashes from entering the eyes. This eyewear fits snugly to the face and usually has baffles that allow air to circulate while protecting against liquid splash. If the baffles are removed or flipped open, the splash protection is compromised. We recommend that you use safety eyewear that provides both chemical-splash and ANSI Z87.1-compliant impact protection. (Chapter 10, pp. 161–162)

Standard Precautions: Procedures that must be followed in order to prevent the transmission of diseases via contact with human body fluids. They incorporate Universal Precautions, which describe methods of preventing the transmission of bloodborne diseases. (Chapter 10, p. 156)

storeroom: A storage room, situated close to the science room, preferably with direct access to the science room via a separately lockable door, in which supplies for the science program can be stored. All chemical stocks should be in storerooms dedicated to and equipped solely for the safe storage of chemicals. Students should never be allowed in chemical storerooms. (Chapter 4, p. 50)

talc: A fine-grained mineral with a soapy feel. Crushed talc was once used to prepare talcum powder, face powder, and some types of short-grained rice. Since the discovery that the mineral is often associated with carcinogenic asbestos fibers, it has largely been replaced by cornstarch.

teratogen: A substance that can cause malformations of an embryo or fetus. Many chemicals, including some frequently used biological stains, can be teratogenic and should be used with great caution. Material safety data sheets (MSDSs) and the Merck Index can be used to check for teratogenic properties of chemicals.

Universal Precautions: Procedures that must be followed in order to prevent the transmission of human bloodborne diseases. The newer Standard Precautions incorporate Universal Precautions. (Chapter 10, p. 156)

UV: Ultraviolet radiation. A component of sunlight that can cause damage to the skin and retina. Excessive exposure to UV radiation may be responsible for an increased risk of skin cancer and cataracts years later. (Chapter 9, p. 141)

video cam: A video camera that can be mounted on a microscope or other optical instrument to send signals to a projection device that permits the whole class to observe an event that might otherwise be too small for everyone to observe simultaneously.

VOIP: Voice over internet protocol. This is a type of telephone service that might not support direct access to 911 emergency services.

zoonotic disease: A disease of animals that can be transmitted to humans.

Appendix A

Chemicals to Go—Candidates for Disposal

The chemicals in the following table were once used in programs and demonstrations. They now are generally considered too hazardous to store and use in precollege programs. You should check the materials you have in your classroom and stockroom. Anything you do not need for your program should be removed. If your program requires any of the chemicals in this table, you should review the hazards involved and make sure you have the expertise, training, facilities, protective equipment, and sufficiently mature students to ensure the education benefits outweigh the risks. Consider replacing these chemicals with a safer alternative.

Most of these chemicals require special handling and special disposal. If you find these, or other hazardous materials, **do not just discard or dump these items.** Some may have decomposed to the extent that even moving or opening the stock bottles could present a serious toxicity or explosion hazard. Consult with professional hazardous waste experts such as those with your state environmental protection agencies.

This is not a list of all the chemicals that might be hazardous. The Merck Index, published and updated regularly by Merck and Company, Inc., is an excellent reference for determining the hazards of chemicals, as is the Flinn Scientific Catalog/Reference Manual, available at *www.flinnsci.com*.

Chemical Name	Chemical Formula	Possible Appearance	Outdated Use	Hazard—Comments
Ammonium dichromate	$(NH_4)_2Cr_2O_7$	Red-orange crystals	Demonstration volcanoes	Unstable; produces toxic by-products when burned
Benedict's solution		Blue liquid	Test for sugar	Caustic; not appropriate below high school level; consider substituting test strips.
Biological stains: Hematoxylin Safranin Methyl orange Methyl red	Various	Strong colors	Slide preparation	Many are dark permanent stains, difficult to remove from skin and clothing; some have been found to be toxic, carcinogenic, or teratogenic (causing malformations in embryos or fetuses of pregnant women); check latest information prior to use
Calcium carbide	CaC_2	Grayish-black lumps	Mixed with water to release acetylene	Fire and explosion hazard
Carbon tetrachloride	CCl_4	Clear, colorless liquid	Organic solvent	Poison by inhalation and skin absorption; carcinogen
Chlordane		Amber-colored liquid	Pesticide	Easily absorbed through skin and mucous membranes; highly toxic; EPA has banned from use
Chloroform	$CHCl_3$	Clear, colorless liquid	Anesthetic	Human inhalation can cause death; EPA lists as carcinogen
Colchicine	$C_{22}H_{25}NO_6$	Liquid	Preparation of mitotic slides	Highly toxic
Concentrated inorganic acids	(e.g., HNO_3, HCl, H_2SO_4, H_3PO_4)	Liquids	Various	Highly corrosive; some are volatile as well; serious burn and eye-damage hazard
Diethyl ether	$C_2H_5OC_2H_5$	Clear liquid (usually in a metal can)	Anesthetizing insects	Forms explosive peroxides slowly on exposure to air
Elemental mercury	Hg	Silver-colored liquid	Thermometers and barometers; illustration of density; electromotive replacement demos	Highly toxic; absorbed through the skin; vapors readily absorbed via respiratory tract

Chemical Name	Chemical Formula	Possible Appearance	Outdated Use	Hazard—Comments
Elemental potassium	K		Demonstrate elements	Forms explosive oxides slowly on exposure to air
Elemental sodium	Na	Grayish non-uniform lumps	Explosive oxidation	Violent reaction with water releasing concentrated NaOH fumes and spray
Formaldehyde solution (also called formalin)	HCHO	Colorless clear or cloudy liquid	Preservative	Strong skin and mucous membrane irritant; carcinogen
Magnesium strips	Mg	Slim silvery coiled metal	Wicks for demo volcanoes; "sparklers"	Burns at very high temperature, releasing UV light that may damage eyes
Mineral talc	$Mg_3Si_4O_{10}(OH)_2$	Soft, white, or gray mineral or white powder	Moh scale mineral; demonstration of cratering	May contain asbestos and cause respiratory problems
Phosphorous, white or yellow	P	White or yellowish waxy-looking sticks stored in water	Demonstrate spontaneous combustion	Spontaneous combustion; fire that is very difficult to extinguish because small particles remaining continue to reignite
Picric acid	2,4,6-trinitrophenol	Clear to yellowish liquid	Specimen preservative	Unstable and explosive if allowed to dry to less than 10% water content
Potassium chlorate	$KClO_3$	White crystals	Generation of oxygen	Strong oxidizer that can cause violent reactions
Potassium cyanide	KCN	White granules or powder	Insect killing jars	Highly toxic; decomposes on exposure to air and moisture to produce deadly hydrogen cyanide gas; can be absorbed through skin or mucous membranes
Silver cyanide	AgCN	White or grayish powder	Silver plating; creating mirrored surfaces	Toxic; can be absorbed through skin or mucous membranes
Sodium hydroxide	NaOH	Small milky-colored pellets	Demonstration of exothermic reactions and acid/base experiments	Highly caustic; serious permanent eye-damage hazard from splash and fumes released during reactions

Appendix B
NSTA Position Statement on Safety

Safety and School Science Instruction

Preamble

Inherent in many instructional settings including science is the potential for injury and possible litigation. These issues can be avoided or reduced by the proper application of a safety plan.

Rationale

High quality science instruction includes laboratory investigations, interactive or demonstration activities and field trips.

Declarations

The National Science Teachers Association recommends that school districts and teachers adhere to the following guidelines:

- School districts must adopt written safety standards, hazardous material management and disposal procedures for chemical and biological wastes. These procedures must meet or exceed the standards adopted by EPA, OSHA and/or appropriate state and local agencies.
- School authorities and teachers share the responsibility of establishing and maintaining safety standards.
- School authorities are responsible for providing safety equipment (i.e., fire extinguishers), personal protective equipment (i.e., eye wash stations, goggles), Material Safety Data Sheets and training appropriate for each science teaching situation.
- School authorities will inform teachers of the nature and limits of liability and tort insurance held by the school district.

- All science teachers must be involved in an established and on-going safety training program relative to the established safety procedures which is updated on an annual basis.
- Teachers shall be notified of individual student heath concerns.
- The maximum number of occupants in a laboratory teaching space shall be based on the following:

  1. the building and fire safety codes;

  2. occupancy load limits;

  3. design of the laboratory teaching facility;

  4. appropriate supervision and the special needs of students.
- Materials intended for human consumption shall not be permitted in any space used for hazardous chemicals and or materials.
- Students and parents will receive written notice of appropriate safety regulations to be followed in science instructional settings.

References

Section 1008.0 Occupant Load—BOCA National Building Code/1996

Section 10-1.7.0 Occupant Load—NFPA Life Safety Code 101-97

40 CFR 260-70 Resource Conservation and Recovery Act (RCRA)

29 CFR 1910.1200 Hazard Communication Standard (Right to Know Law)

29 CFR 1910.1450 Laboratory Standard, Part Q The Laboratory Standard (Chemical Hygiene Law)

National Research Council (1995). *Prudent Practices in the Laboratory*, National Academy Press.

Furr, K. Ed. (1995). *Handbook of Laboratory Safety*, 4th Ed. CRC Press.

Fleming, et al Eds. (1995). *Laboratory Safety*, 2nd Ed. ASM Press.

National Science Education Leadership Position Paper. (1997). Class size in laboratory rooms. *The Navigator.* 33(2).

Authors

George R. Hague, Jr., Chair, Science Safety Advisory Board, St. Mark's School of Texas, Dallas, TX 75230

Douglas Mandt, Immediate Past-Chair, Science Safety Advisory Board, Science Education Consultant, Edgewood, WA 98372

Dennis D. Bromley, Safety Instructor, Independent Contractor, Anchorage, AK 99502

Donna M. Brown, Radnor Township School District, Wayne, PA 19087

Frances S. Hess, Cooperstown H.S., Cooperstown, NY 13326

Lorraine Jones, Kirby H.S., Nashville, TN

William F. McComas, Director, NSTA District XVI, University of Southern California, Los Angeles, CA 90089

Kenneth Roy, Glastonbury Public Schools, Glastonbury, CT 06033

Linda D. Sinclair, South Carolina Department of Education, Columbia, SC 29201

Colette Skinner, Henderson, NV 89015

Olivia C. Swinton, Patricia Roberts Harris Education Center, Washington, D.C.

Nina Visconti-Phillips, Assistance & Resources Integrating Science Education (ARISE) Dayton, NJ 08810

—Adopted by the NSTA Board of Directors, July 2000

Index

Note: Page numbers in *italic type* refer to information in sidebars or tables.

Index

Index

Index

Index

Index

Index

Index

T

Index